Reflections on Cultural Heritage Theories and Practices

A series by the Raymond Lemaire International Centre for Conservation

ISSN: 2576-3075
eISSN: 2576-3083

Volume 3

EDITED CONTRIBUTIONS TO THE INTERNATIONAL CONFERENCE ON INNOVATIVE BUILT HERITAGE MODELS AND PREVENTIVE SYSTEMS (CHANGES), 6–8 FEBRUARY 2017, LEUVEN, BELGIUM

Innovative Built Heritage Models

Editors

Koen Van Balen & Aziliz Vandesande

Department of Civil Engineering and Raymond Lemaire International Centre for Conservation, KU Leuven, Heverlee, Belgium

CRC Press
Taylor & Francis Group
Boca Raton London New York Leiden

CRC Press is an imprint of the
Taylor & Francis Group, an **informa** business
A BALKEMA BOOK

Cover illustration: Window maintenance workshop "JANELAS com TINTA têm muito + PINTA" in historic Porto © Teresa Cunha Ferreira

CRC Press/Balkema is an imprint of the Taylor & Francis Group, an informa business

Typeset by V Publishing Solutions Pvt Ltd., Chennai, India
Printed and bound in Great Britain by CPI Group (UK) Ltd, Croydon, CR0 4YY

Published by: CRC Press/Balkema
Schipholweg 107C, 2316 XC Leiden, The Netherlands
e-mail: Pub.NL@taylorandfrancis.com
www.crcpress.com – www.taylorandfrancis.com

ISBN: 978-1-138-49861-7 (Hardback)
ISBN: 978-1-351-01479-3 (eBook)

Innovative Built Heritage Models – Van Balen & Vandesande (Eds)
© 2018 Taylor & Francis Group, London, ISBN 978-1-138-49861-7

Table of contents

Innovative Built Heritage Models – Van Balen & Vandesande (Eds)
© 2018 Taylor & Francis Group, London, ISBN 978-1-138-49861-7

Committees

ORGANISING COMMITTEE

Prof. Koen Van Balen, *KU Leuven University, Belgium*
Dr. Aziliz Vandesande, *KU Leuven University, Belgium*
Prof. Stefano Della Torre, *Polytechnic University of Milan, Italy*
Prof. Silvia Naldini, *Delft University of Technology, The Netherlands*
Prof. Christer Gustafsson, *Uppsala University, Sweden*
Eng. Dario Foppoli, *Foppoli Moretta e Associati Consulting Engineers, Italy*

INTERNATIONAL SCIENTIFIC COMMITTEE

Stefania Bossi, *Italy*
Gabriella Garcia Velez, *Ecuador*
Jan Haenraets, *USA*
Herdis Heinemann, *The Netherlands*
Terje Nypan, *Norway*
Eduardo Rojas, *Chile/USA*
Nathalie Van Roy, *Belgium*
Maarten Vergauwen, *Belgium*
Luc Verpoest, *Belgium*

Introduction

Innovative Built Heritage Models – Van Balen & Vandesande (Eds)
© 2018 Taylor & Francis Group, London, ISBN 978-1-138-49861-7

Innovative built heritage models based on preventive and systemic approaches

A. Vandesande & K. Van Balen
Raymond Lemaire International Centre for Conservation and Department of Civil Engineering, KU Leuven, Heverlee, Belgium

1 INTRODUCTION

The research presented in this volume is linked with two of the most challenging and discussed changes that our society is facing today, i.e. ongoing environmental sustainability concerns and the early 2010s general economic decline's consequences. In this context, renovation and refurbishment of the existing building stock is becoming a dominant aspect of the construction industry, whereby maintaining existing structures contributes to i.a. environmental viability (Power 2010), prolongation of the physical service life of buildings and building parts (Thomson and van der Flier 2009), waste-avoidance activity and preserving embodied energy (Bullen and Love 2010).

On the one hand, this transition can benefit from knowledge gained in the built heritage conservation field. The strategic research agenda of the European Construction Technology Platform states that European construction industry will achieve greater competitiveness and the ability to satisfy societal needs through research, development and innovations oriented towards protecting and maintaining cultural heritage, and (adaptive) re-use of existing buildings (ECTP 2008). On the other hand, these changes are also finding resonance in the built heritage sector, as the over-proportional investment of economic, societal and other resources in large restoration or regeneration works sometimes leads to the deterioration of the older part of the building stock and individual buildings that are not sufficiently maintained to preserve their value (Hassler 2009). In addition, built heritage is increasingly accepted as 'ingrained' in the urban and rural environment, therefor it will have to adapt to unique challenges and uncontrolled change to increasing nature and human driven risks (Wilby 2007, UNHABITAT 2011).

At the same time, the importance of preserving and managing these assets ingrained in the contemporary urban and rural fabric for future generations is increasingly acknowledged. Since the early 2010s, several international and European policies strongly stress the potential of heritage in the sustainable development discourse. Whereas UNESCO and the Council of Europe are prominent actors in this evolution, more recently 2 European Union institutions in effect strengthened this new discourse (Vandesande and Van Balen 2015), i.e. the Council of the European Union by adopting the 'Conclusions on cultural heritage as a strategic resource for a sustainable Europe' (Council of the European Union 2014) and the European Commission by adopting the 'Communication on an integrated approach to cultural heritage in Europe' (European Commission 2014).

This trend reflects the importance given to sustainable approaches to our existing building stock and the role of built heritage in this process. Consequently, sometimes arguments rise against large restoration works that can potentially result in postponed interventions, i.e. the high cost of restoration, the always-increasing demand for funding and the uncertainty of sufficient revenue to cover costs of property management. In response, preventive conservation, maintenance and monitoring are gradually recognised by researchers given its effectiveness in site management and its long-term beneficial effect on physical authenticity and integrity of buildings. Thereby the main benefits of a preventive conservation approach for built heritage are considered the cost-effectiveness for built heritage owner-managers (Forster and Kayan 2009), the quality protection of built heritage and environmental enhancement (Vandesande and Van Balen 2016), the inclusion of the heritage sector in employment policies (Iżvolt 2015) and the empowerment of local communities in dealing with heritage (Della Torre 2010).

However, ongoing research in context of the UNESCO chair on preventive conservation, monitoring and maintenance of monuments and sites (PRECOM³OS), established at the Raymond Lemaire International Centre for Conservation at KU Leuven University, demonstrates that in practice only few Preventive Conservation (PC) process case studies are operational in the built heritage sector. Although a pro-active approach towards monitoring gained importance since the 1990s, as 'a reflection of the growing commitment to improving management frameworks for care of cultural heritage through the use of monitoring, which is understood as a key component of the management process' (Stovel 2008: 15), the built heritage filed has largely failed to resolve the issue of maintenance and in extension of preventive conservation. It has been stated that 'very little has been written specifically dealing with systematic maintenance of or for building conservation' and that 'there is no well-established academic study or research to ascertain why maintenance is not widely practiced, disseminated and developed' (i.a. Worthing et al. 2002:292; Dann and Cantell 2005:42; Forster and Kayan 2009:211).

A concise literature review indicates that this issue is most commonly attributed to maintenance market failure (Fig. 1). Some authors note that building owners have negative perceptions associated with maintenance (Hills and Worthing 2006) and perceive it as a 'difficult burden' (Feilden 2003: 238), while other researchers are oriented more towards the lack of maintenance skills (Hutchings and Corr 2012) or the reasons contractors are not interested in the 'cinderella sector' of construction industry and low-status professional maintenance operations (Wood 2005: 291).

Moreover, built heritage maintenance entails a service which in itself cannot be effectively traded in terms of strategic market management, i.e. the service is not an identical, interchangeable and standardised commodity; it is not yet accepted in the market and the demand and supply side do not have sufficient price information and assurance (Sanchez and Heene 2004).

Therefore, the Raymond Lemaire International Centre for Conservation (KU Leuven University) considered 'CHANGES' a relevant and timely topic for its yearly international conference, the 'thematic week'. This edition twins with the 'CHANGES' project, supported by the Joint Programming Initiative Heritage Plus programme. CHANGES

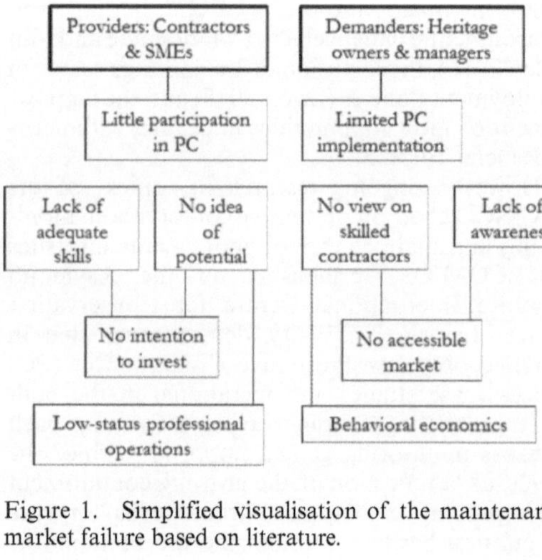

Figure 1. Simplified visualisation of the maintenance market failure based on literature.

stands for Changes in Cultural Heritage Activities: New Goals for Economy and Society. The main goal of this European collaboration is to demonstrate the effectiveness of Preventive and Planned Conservation, Maintenance and Monitoring by an increased understanding of conservation and valorisation as preventive measures, the effectiveness of maintenance and economic mechanisms underlying built heritage and construction sector. The scientific consortium of the conference consisted of members representing the different CHANGES consortium organisations, i.e. Politecnico di Milano (Department of Architecture, Built Environment and Construction Engineering) as lead partner, the University of Leuven (RLICC and Department of Civil Engineering), TU Delft (Faculty of Architecture and the Built Environment), Uppsala University (Department of Art History) and Foppoli Moretta e associati consulting engineers.

This volume specifically reports on the lectures and fruitful debates on heritage impact during the thematic week, which took place February 6th-8th 2017. The conference's aim was to provide an international overview of the few existing strategies, processes and operational case studies that support the implementation of a preventive and planned conservation approach in the built heritage sector. Based on common gaps and re-occurring tendencies, the findings of the contributions in this volume can meet the increasing demand for shared insights and formation and support the transition towards a more sustainable conservation process. This publication very straightforwardly reflects on CHANGES by providing research, case studies and reflections that can serve as baseline records, guidance-and hopefully inspiration. The volume consists of three main parts which are shortly outlined below: 'The CHANGES paradigm', 'Preventive and planned conservation' and 'Case studies of Change'.

2 THE CHANGES PARADIGM

The first chapter not only entails a critical review of approaches for built heritage which provide opportunities for changes, but also provides a broader perspective with specific frameworks and mechanisms on different scales. The different contributions—each from their own perspective—offer an understanding of the complexity of the subject, identify what is known and seek new lines of enquiry.

The keynote by **Stefano Della Torre** proposes to induce changes in the built heritage sector through introducing programming in the overall conservation process. The author puts forth that today there is a multitude of best practices and effective tools available to implement more preventive and planned approaches in the built heritage sector. However, these are not often considering the conservation process as whole, thus including the overall management process, facility management, conservation works and sub-processes. To achieve changes in current practice, the author stresses the importance of programming and planning. Interventions should not be simply designed and executed, but they should be carefully prepared in order to guarantee a set of benefits on the long run. In this vision, the alternative between preventive and curative approaches is not absolute, but the output of logical decisions based on management control, risk assessment and maintenance systems based on regular and structural inspections. Given the diversity of the hazards at different scales, changes also implies a vision that goes far beyond the borders of a single property. The paper provides several arguments to move towards a more systemic vision which takes into account interoperability, long-term planning, a link with the surrounding environment and local economy. Herewith, it is concluded that programming should ensure that resources are not wasted on inefficient interventions.

Essentially, the keynote calls for turning interventions on built heritage into opportunities for changes in the direction of implementing more preventive conservation based tools leading to sustainability, resilience and cost-effectiveness. Within the chapter, several practical frameworks and mechanisms which can contribute to this change are discussed, both on the level of existing buildings and cities. On the level of individual structures, **Marco Morandotti** introduces a resilience evaluation method which assesses the impact of re-use interventions based on the expected impact by users and the maximum rate of transformation with reversibility. Similarly, **Simone Lucenti** provides a review of standards European historic universities to highlight current trends which require the reduction of office space, the growing need to improve the efficiency of available spaces and ensuring the maximum compatibility between function and space. On the level of historic cities, **Maria Siguencia Avila and Silvana Vintimilla Andrade** give a practical analysis of the 2011 UNESCO Recommendation on Historic Urban Landscapes (HUL) and its six-step action plan to achieve beyond-buildings heritage conservation. Their comparative analysis of 15 HUL based initiatives in Latin America provides several insights of successful practices and their possible improvement. Within the scope of implementing changes through built heritage conservation, the observation that the lack of long-term visions and maintenance correlates with loss of historic structures and increased gentrification is especially relevant. The authors also conclude that the effectiveness of change is largely determined by the establishment of clear roles of stakeholders. A similar observation was made by **Cristina Boniotti** in her contribution on change management for built heritage through public private partnership. Given that recent actions strongly promote cross-sectoral collaboration for built heritage and European directives are encouraging the participation of private resources in valorisation interventions, the author stresses the need to promote collaboration among local actors who are not used to cooperating in the sector.

Next to the observations on cross-sectoral collaborations, there is also a need to define roles and procedures to kick-start changes for all usual suspects involved in built heritage conservation. **Stefania Bossi** gives a highly relevant analysis on the participation of the restoration companies in the built heritage conservation process. Based on ten years of monitoring the strengths and weaknesses of existing strategies and interviews with involved shareholders, the author makes several observations on actual tendencies in the built heritage market, the corrections needed and the operative role of the restoration companies herewith.

3 PREVENTIVE AND PLANNED CONSERVATION

The initial seed to research on preventive conservation was planted by Raymond Lemaire (1921–1997), founder of the Raymond Lemaire International Centre (RLICC) at the University of Leuven, official reporter at the Second International Congress of Architects and Specialists of Historic Buildings in Venice (1964) and co-author with Piero Gazzola (1908–1979) and Roberto Pane (1897–1987) of the final Venice Charter. During a radio interview in 1975, on the occasion of the European Architectural Heritage Year, he stated that ‘we are too easily thinking today that always an old building must be restored. That is not quite so (…). Buildings are being restored that do not need restoration at all (…). Let us first and foremost conserve buildings. And to do that, we need maintenance’ (Fonds Raymond Lemaire 1921–1997: 3401). However, the actual research and practice required to implement preventive conservation approaches dealing specifically with built heritage was only triggered during the last 3 decades. Since the mid 1990s, the topic became increasingly endorsed within international and European policies, which in turn was also reflected in different EU-funded research projects and by international organisations that started implementing a pro-active approach towards monitoring and maintenance, such as the World Monument Fund with the Watch Programme and the UNESCO with the Reactive Monitoring and Periodic Reporting mechanisms. In line with the evolutions and shifts that characterise preventive and planned conservation today, the authors in this chapter provide contributions on its actual scope, efficiency evaluation and new opportunities for its implementation.

The keynote by **Aziliz Vandesande and Koen Van Balen** observes that preventive conservation of built heritage is increasingly of interest to academic research and professionals in the sector, with promising examples that demonstrate the viability of this approach. Accordingly, there is ample data and material from different research fields, with scopes and focus. Nevertheless, a general overview or commonly held delineation is not yet developed. Therefore the authors develop a working definition for preventive conservation of built heritage which starts from condition-based maintenance with a monitoring system. The paper puts the working definition forth as the least destructive approach to interventions which are inevitably necessary to maintain a building's functional, structural and technical performances and extend the life-cycle of built heritage values, authenticity and integrity. While this preventive conservation approach has several benefits, the authors observe that its practice and implementation are characterised by different influencing factors, i.e. the available knowledge and know-how as well as financial, environmental and social factors. They conclude that these influencing factors could explain the observation that maintenance management of public built assets gains a significant importance, but the interest of built heritage field in preventive conservation is lagging back.

In addition to defining what preventive and planned conservation implies for built heritage, other contributions focus on learning lessons from the commonly associated field of risk management. **Anthoula Konsta** provides a review of recent actions, including recommendations, guidelines, best practices and case studies to highlight specific components of risk management methodologies that should be integrated preventive and planned conservation strategies for built heritage. In contrast, **Lorenzo Cantini** starts from a more specific point of view, by focusing on preventive actions for built heritage against seismic risk. The paper combines findings from past and present preventive actions to tackle the problem of seismic risk of built heritage and the most recent Italian guidelines for the mitigation of the seismic risk. The author concludes that there are several preventive measures which can be implemented for built heritage, both in terms of mitigation and adaption, e.g. increase the quality of the construction techniques. However, currently these measures are mostly used for public strategic buildings or iconic sites and only rarely applied to the larger built heritage stock. Herewith a strong case is made for continuous care by monitoring built heritage to guarantee the reliability of the interventions and efficient data collection on the structural safety in post disaster context.

In order to increase the efficiency of data collection and facilitate the processing of monitoring data, several innovative technology applications are introduced to support preventive and planned conservation approaches on different scales. On level of single structures, **Stefano Della Torre, Rossella Moioli and Alessandra Pili** discuss the challenges of moving from interventions as an event to interventions as a continuous process. Herewith, it is argued that the use of a relational database is more efficient in organising the conservation plan and process of complex built heritage structures. Based on a practical application of a platform based on a relational database, the Planet-Beni Architettonici, they demonstrate its usefulness in terms of conservation management and optimising financial and human resources. On the larger scale of historic urban areas, **Olga Zalamea, Jos Van Orshoven and Thérèse Steenberghen** introduce the application of knowledge-based representations for built heritage. In its most basic form, knowledge-based representations are the field of artificial intelligence dedicated to representing information about the world in a form that a computer system can utilize to solve complex tasks. The authors discuss how they identified knowledge-based representations for built heritage and their strengths and weaknesses in supporting the development of a preventive conservation system. The use of these representations can provide a solution to deal with the complex data integration problem of built heritage management whereby multiple actors from various disciplines use a variety of data sources and data models. The main focus of the paper is explicitly representing 3D features by using the City Geography Markup Language (CityGML) standard, combined with an ontological approach to achieve the semantic refinements required for built heritage. These refinements include thematic concepts, such as vulnerability or legal protection issues, as well as built components, such as columns or cornices, which are tackled by using 2 pre-existing ontologies: the Geneva-CityGML ontology and MONDIS (the monument damage ontology).

Another topic dealt with in the chapter on preventive and planned conservation is the evaluation of existing monitoring and condition-based maintenance systems, namely the Monumentenwachten or Monument Watches (MOWA). The latter are experienced and highly professional organisations that implement a preventive conservation process. Their central rationale is timely identification and limited correction of defects on historic structures to reduce deterioration of the fabric and prevent major consequential damage. In practice, their core activity is supporting private and public built heritage owner-managers in the preservation of their properties through a periodic monitoring system. The MOWA model finds it origination in The Netherlands during the early 1970s and was later on established in Belgium, more specifically in Flanders in 1991. Since the 2000 Council of Europe campaign "Europe, a common heritage", MOWAs are increasingly known in various European countries and regions. Other organisations that follow this model have been implemented in i.a. the UK (Maintain Our Heritage in Bath), Denmark (Bygningsbevaring) and Germany (Denkmalwacht). More recently there are other establishing attempts, i.e. Műemlékőr in Hungary and the Traditional Buildings Health Check Scheme in Scotland (Vandesande and Van Balen 2016).

A first contribution by **Herdis Heinemann and Silvia Naldini** provides a very clear understanding of the long-term impact of a MOWA organisation. Based on a qualitative multiple case study research, the evolution of modus operandi and role of MOWA in The Netherlands over a period of 40 years is analysed. The authors observe that MOWA started as an organisation that supports owner-managers of built heritage by signalling damage, but due to its close relation to owner-managers has evolved to the main advisor in maintenance interventions. By means of analysing 16 case studies in terms of type of ownership, building typology and state of conservation, the authors conclude that MOWA as main advisor and shortcut to approaching a contractor can lead to acceptable results. In addition, **Silvia Naldini, Herdis Heinemann and Rob van Hees** apply a Failure Mode and Effects Analysis (FMEA) to provide a qualitative indication of the positive effects of preventive and planned conservation within the scope of MOWA in The Netherlands. Next to understanding the role and positive effects, **Nathalie Van Roy et al.** analyse the effect of MOWA in Flanders on the maintenance behaviour of owner-managers of built heritage. In order to identify maintenance practices and to understand the types of interventions that are executed, the authors applied a mixed method approach which combines social research methods, a technical analysis of the performed interventions and a statistical analysis of the inspection reports of MOWA. The results indicate that the owners take the severity of damages that are detected by MOWA into account when they decide on maintenance actions, but that their actions are also influenced by practical considerations, i.e. the accessibility of the location of damage, the clustering of interventions, the availability of funding, the lack of quality control during the execution of works, the durability of the initial design choices and the availability, knowledge and skills of the contractor.

Most contributions in the chapter on preventive and planned conservation deal with the lack of implementing preventive actions in the built heritage sector. Similarly, **Luís Ramos et al.** observe the lack of a systemic preventive conservation policy of built heritage in South-West Europe, which in daily practice is translated into unplanned and overpriced approaches for inspection, diagnosis, monitoring and interventions. In order to tackle this situation, the authors present a highly relevant and innovative project, HeritageCARE, which incorporates several findings of the contributions in this chapter. In short, the project envisages develop a 4D database for buildings/assets management, information exchange, and the standardization of methods and tools for maintenance and preventive conservation of heritage buildings/assets. The database is envisaged to have 3 service levels, ranging from to rapid condition screening to Building Information Modelling (BIM) for heritage structures, to ensure adaptability to owners' needs and financial resources. Whereas the project directly involves society, universities, private sector and public institutions, the project aims to simultaneously raise awareness on the benefits of preventive and planned conservation and hopes to create a self-sustaining entity.

4 CASE STUDIES OF CHANGE

Whereas case studies have always been considered 'particularly well suited to new research areas or research areas for which existing theory seems inadequate' (Eisenhardt, 1989: 548), this chapter provides a selection of innovative built heritage conservation practices that induce specific changes in local environments or societies and which hold a strong link with preventive and planned conservation.

The keynote by **Christian Ost and Francis Carpentier** frames this chapter by introducing some more established concepts on conservation dynamics. Based on the rapid speed of policy changes related to cultural heritage and the continuous changes in the economic climate, the authors propose a paradigm shift which intrinsically links the cultural, social, and economic needs of contemporary society to the conservation of built heritage. In line with this paradigm shift, the authors put forth that through specific mechanisms such as community involvement and the use of local resources and skills, cultural heritage conservation can play an important role in the transformation of society and in changes towards sustainable solutions.

The case studies presented in this chapter each demonstrate that this paradigm shift is also practically applied by built heritage professionals. The first contribution by **Rossella Moioli** is a very literal example of change as it deals with the progressive re-use of and obsolete historical structure, a spinning mill in Sulbiate, Italy. By means of combining an innovative built heritage conservation project and long-term management model, the case study demonstrates very well that built heritage can be a public resource and the effectiveness of a well-planned conservation process. The Sulbiate project is specifically interesting as it directly involved local public and private actors in a partnership and in the formulation of the design process, creating a local learning laboratory.

In contrast to this very literal example of change through re-use, **Christer Gustafsson and Akram Ijla** introduce a very operational and on multiple occasions demonstrated model for creating changes at a regional level. They discuss the Halland model as an approach to activate building conservation as a catalyst for regional sustainable development. The model was named after a regional joint venture in the region of Halland in Sweden. During 1990s recession it aimed at increasing the total volume of construction and to absorb the unemployed construction workers within the projects' activities. The main mantra of the model remains until today: *save the jobs, save the craftsmanship, save the buildings*. Based on a local network of the construction industry, the historic built environment sector, real estate owners, local and regional authorities and trade unions the venture soon developed into a cross-sectoral action for workforce development. Today the model reflects a very specific approach of collective governance aiming at planning and managing sustainable long-term actions for built heritage conservation. Within the framework of the social economy sector, unemployed construction workers and apprentices were trained in traditional building techniques. During this training they gained proficiency on historic buildings at risk under the supervision of skilled craftsmen and conservation officers, which holds benefits for different sectors. In the process, historic buildings at risk were saved from demolition, a younger generation is trained in craftsmanship and new jobs were created. Very similar mechanisms can today be observed throughout different regions in Europe, among others in the Valtellina Cultural District in Italy. This specific case study presented by **Dario Foppoli** was initiated in 2010 and is part of the larger *Distretti Culturali* project co-funded by the Cariplo Foundation. The scope of the project is wide, including work packages aimed at restoring heritage buildings, Preventive Planned Conservation, skills building, community involvement and valorisation. Within the project, the activities dealing with the "The Route of Terraces" proved most suitable for creating networks and stimulating the link between local population and cultural landscape. Based on the weaknesses and threats inherent to the territory, the "+++ More Positive Signs" initiative was developed by a network of both public entities and social cooperatives to ensure the maintenance of paths, terraces and landscape in the Valtellina district. The project brings together the weaknesses of mountain territories, including both social and hydro-geological weaknesses, and the need to create local employment.

One of the central components of these case studies aimed at change through built heritage conservation is the very active involvement of local actors. This specific mechanism is highlighted in the paper presented by **Teresa Cunha Ferreira** which deals with the set-up of a preventive and planned conservation system. The case study entails a very promising Portuguese experiment, the Romanesque Route, which demonstrates that preventive conservation is increasingly successful when linked with the empowerment of local communities and users. Throughout the project life-span, several additional were created such as employment in small local contractor firms, the qualification of local users for the future safeguarding and sustainability of their heritage and a sharp decrease in economic and ecological resources.

5 LESSONS LEARNED

The different contributions in this volume point out that the over-proportional investment of financial and societal resources in large, often prestigious, built heritage interventions without considering a long term-perspective, quality protection, programming and post-conservation management has become unsustainable in many different ways. Thereby, several research outcomes highlight the importance of involving local end-users, communities and traditions—either when dealing with individual properties or the larger part of the historic urban or rural environment—to ensure that actual changes can be implemented in local systems. Taking full potential of all these factors, combined with practical approaches that link built heritage conservation interventions to needs of contemporary society, is a successful formula to initiate changes on a local scale.

This observation becomes especially clear in the different case studies of change, whereby the conservation of built heritage is no longer the starting point of a local initiative but completely ingrained in cross-sectoral collaborations aimed at sustainable development. Indeed, already in 1992, Agenda 21 emphasised the local level for meeting sustainability challenges and called for local participatory approaches for sustainable urban development and the strengthening of the capacity of governing bodies. Since the 2010s, several international organisations stress the potential of medium-sized cities, public places and street as drivers of urban prosperity including: environmental sustainability, quality of life and social inclusion (to name some: UNESCO 2011, UN HABITAT 2013). Similarly, the research field has demonstrated that innovative public policies joining sustainable development and historic city-regions are mostly driven by local public authorities (Roders and Van Oers 2011) and that local specialised services geared for specific activities are more efficient and provide space for learning processes in relation to user preferences, regulations and identifying market openings (Turok 2010). In line with this, several research fields and disciplines dealing with innovations focus on the local level and denominate their geographical working space accordingly, i.a. the 'field' and its respective 'habitus' with its own specific logic, currency and goals, 'transition arenas' as specific social environments which enable sustainable development (Loorbach 2007), 'urban transition labs' as a locus of problems and innovations (Nevens et al. 2013) and 'niche space' or experimental pilot projects to build social networks (Frantzeskaki et al. 2012).

In considering local as a proxy for development, a distinction should be made between local networks and communities versus geographical proximity. Herewith, local networks are directly related to local interests, organisational contexts, sense of community, social cohesion and emotional attachment. Proximity is tied to agglomeration economies, which essentially are geographical concentrations of a market that cannot be explained by the presence of natural resources. Within the scope of CHANGES, the different contributions demonstrate that both are crucial, as built heritage is in itself geographically immovable and therefore its management is ingrained in local networks and communities. This local proxy for built heritage provides several components that contribute to the sustainability of its management and development opportunities.

While an increasing interest has recently been shown for projects planned and carried out by local actors and related to specific areas, eventually CHANGES should also move from small scale projects to more systemic and fundamental changes in the built heritage sector. This is especially relevant as environmental and economic concerns are driving investments in the construction sector towards more sustainable approaches. From one hand many actors in the construction overlook the contribution of cultural heritage in these approaches, from the other hand the built heritage field is lagging back in preventive conservation. In response, several authors propose a link with the well-established field of business strategies to enable the upscaling of preventive and planned conservation in the built heritage sector.

REFERENCES

Bullen, P. & Love, P. (2010). 'The Rhetoric of adaptive Reuse or Reality of Demolition'. Cities, 27(4): 215–224.

Council of the European Union. (2014). Conclusions on Cultural Heritage as a Strategic Resource for a Sustainable Europe, adopted at the Education, youth, culture and sport Council meeting on 20/05/2014.

Dann, N. & Cantell, T. (2005). 'Maintenance: from philosophy to practice', Journal of Architectural Conservation, 11 (1): 42–54.

Della Torre, S. (2010), 'Preventiva, integrata, programmata: le logiche coevolutive della conservazione'. In: Biscontin, G., Driussi, G. (Eds.) Pensare la prevenzione: manufatti, usi, ambienti, Atti del convegno, Venice: Arcadia Ricerche: 67–76.

ECTP. (2008). Vision 2030 & Strategic Research Agenda Focus Area Cultural Heritage: European Construction Technology Platform.

Eisenhardt, K. (1989). Building theories from case study research. Academy of Management Review, 14(4), 532–550.

European Commission. (2014). Communication from the Commission to the European Parliament, the Council, the European Economic and Social Committee and The Committee of the Regions Towards an Integrated Approach to Cultural Heritage for Europe, adopted on 22/07/2014: (COM(2014) 477 final).

Feilden, B. 2003. Conservation of historic buildings. Oxford (Butterworth-Heinemann).

Fonds Raymond Lemaire (1921–1997), Radio interview 1975 on the occasion of the European Architectural Heritage Year, In: University archive KU Leuven (BE / 212934 / P 225), ODIS—Database Intermediary Structures Flanders: Record no. 3401.

Forster, A. & Kayan, B. (2009). 'Maintenance for historic buildings: a current perspective', Structural Survey, 27(3): 210–229.

Frantzeskaki, N., Loorbach & D., Meadowcroft, J. (2012). 'Governing societal transitions to sustainability'. International Journal of Sustainable Development, 15(1/2): 19–36.

Hassler, U., (2009). 'Long-term building stock survival and intergenerational management: the role of institutional regimes'. Building Research & Information, 37(5–6), pp. 552–568.

Hills, S. & Worthing, D. (2006). 'Private home, public cultural asset: the maintenance behaviour of listed building owner-occupiers', Journal of Housing and the Built Environment, 21(2): 203–213.

Hutchings, J. & Corr, S. (2012). 'A framework for access to the conservation-restoration profession via the mapping of its specialist competencies', Higher Education, 63: 439–454.

Iževolt, P. (2015). 'Preservation of the cultural heritage through inclusion of unemployed in Slovakia. Present challenges of conservation practice'. In: Van Balen, K., Vandesande, A. (Eds). Heritage counts, Antwerp: Garant: 305–316.

Loorbach, D. (2007). Transition management: new mode of governance for sustainable development, PhD dissertation, Rotterdam: Erasmus Universiteit Rotterdam.

Nevens, F., Frantzeskaki, N., Gorissen, L. & Loorbach, D. (2013). 'Urban Transition Labs: co-creating transformative action for sustainable cities'. Journal of Cleaner Production, 50: 111–122.

Power, A. (2008). 'Does demolition or refurbishment of old and inefficient homes help to increase our environmental, social and economic viability?'. Energy Policy, 36(12): 4487–4501.

Roders, A. & van Oers, R. (2011). 'Bridging cultural heritage and sustainable development'. Journal of Cultural Heritage Management and Sustainable Development, 1(1): 5–14.

Sanchez, R. & Heene, A. (2004). The new strategic management: organization, competition and competence, New York: Wiley & Sons.

Stovel, H. (2004). 'An Advisory Body View of the Development of Monitoring for World Cultural Heritage'. World Heritage Papers, 10, Paris: UNESCO: 17–21.

Stovel, H. 2008. 'Origins and Influence of the Nara Document on Authenticity'. Association for Preservation Technology Bulletin, 39 (2/3): 9–17.

Thomsen, A. & K. van der Flier. (2009). 'Replacement or renovation of dwellings: the relevance of a more sustainable approach.' Building Research & Information, 37(5–6): 649–659.

Turok, I. (2004). 'Cities, Regions and Competitiveness'. Regional Studies, 38(9): 1069–1083.

UN-HABITAT. (2011). Cities and Climate Change: Global Report on Human Settlements, London: Earthscan Publications Ltd.

UNESCO. (2011). Recommendation on the historic urban landscape, adopted on 10/11/2011.

UN HABITAT. (2013). Streets as public spaces and drivers of urban prosperity: United Nations Human Settlements Programme.

Vandesande, A. & Van Balen, B. (2015). 'Towards a Sustainable Development Rationale for Heritage Management and Research'. In: Van Balen, K., Vandesande, A. (Eds). Heritage Counts, Antwerp: Garant: 11–20.

Vandesande, A. & Van Balen, K. (2016). An operational preventive conservation system based on the Monumentenwacht mode. In: Van Balen, K. & Verstrynge, E. (eds.) SAHC 2016 Structural Analysis of Historic Constructions. Anamnesis, diagnosis, therapy, controls. Leiden: (CRC Balkema / Taylor & Francis Group.

Wilby, R. (2007). 'A review of climate change impacts on the built environment'. Built Environment, 33(1), 31–45.

Wood, B. (2005). 'Towards innovative building maintenance', Structural Survey, 23(4): 291–297.

Worthing, D., Dann, N. & Bond, S. (2002). 'Issues in conservation management', Proceedings of the international council for research and innovation in building and construction global symposium: Applying and Extending the Global Knowledge Base—Working Commission 070 Facilities Management and Maintenance, Glasgow: CIB and CABER: 292–302.

Worthing, D., Dann, N. & Bond, S. (2002). 'Issues in conservation management', Proceedings of the international council for research and innovation in building and construction global symposium: Applying and Extending the Global Knowledge Base—Working Commission 070 Facilities Management and Maintenance, Glasgow: CIB and CABER: 292–302.

CHANGES, the paradigm

Innovative Built Heritage Models – Van Balen & Vandesande (Eds)
© 2018 Taylor & Francis Group, London, ISBN 978-1-138-49861-7

The management process for built cultural heritage: Preventive systems and decision making

S. Della Torre
ABC Department, Politecnico di Milano, Italy

ABSTRACT: Innovation processes in heritage field seem to be less spontaneous than induced by some actors, who identify problems and push towards new solutions. The availability of best practices and effective tools for maintenance and preventive conservation poses new problems concerning the links among overall management process, which includes facility management as well as major conservation works, and sub-processes, which could be thought as standing alone, or inside the frame of a general set of coordinated activities. The paper aims at providing tentative schemes to clarify the possible roles of preventive systems in the frame of overall process. Given the complex and hazardous relationships of built cultural heritage management with the surrounding environment, and its role for local economy, the analysis has to take many factors into account.

1 INTRODUCTION

The understanding of Preventive Conservation and Planned Conservation evolved during the last decade, from a narrow model focused on maintenance to a more ambitious focus on the consistency between all activities concerning cultural heritage.

The first theses about the risks of over-restoration and the benefits of regular inspections and preventative maintenance have been encompassed in more comprehensive models, based on people involvement, knowledge management, care for the impacts, in order to frame cultural heritage into a broader discourse on sustainability (Van Balen & Vandesande 2013; Vandesande, Moioli & Van Balen 2014).

Italian law enforced in 2004 had already go further, as it underscores consistency, coordination and planning as the conditions to make conservation happen. In other words, conservation would not happen if it is not "planned", and all conservation activities, such as investigation, prevention, maintenance and restoration itself should be consistent, coordinated and planned. This vision challenges any contraposition among the activities of continuous care and major restoration/conservation works, focused on curative conservation and restorative treatments, as well as on adaptive changes, which happen after an interval of some years or decades. Why and how major works should be inserted into the same logical frame of a planned, and therefore preventive, conservation?

The often observed lack of coordination between conservation and valorisation of historic buildings is one more conceptual knot: the discourse on costs and investments related to built cultural heritage is often split into two separate balances, as if valorisation activities could be thought and developed separately from conservation issues and contents, and vice versa.

These issues moved the research direction to the program phase, as the one in which the conditions are produced for a better management throughout the life of cultural properties, including preventive schemes, dealing with use, opening to the public, daily care and extraordinary affairs.

2 HYPOTHESES ABOUT CONSERVATION AND VALORIZATION OF BUILT CULTURAL HERITAGE

In order to provide tentative schemes clarifying the possible roles of preventive systems in the frame of the overall process, it is necessary to recall some hypotheses.

A first point is that conservation is not a purpose in itself: the aim is to make cultural landscapes and historic buildings accessible by the broadest public of today and of tomorrow. Accessible means usable, and this is a critical step, as surely use threatens conservation because of "wear and tear" processes. But (historic) buildings have to be used (inhabited, visited), as architecture can be acknowledged and understood only through use, not by sight only (Bellini 1998).

Use is often re-use: the functions change through time, and the permanence of the same functions often is not a guarantee against the obsolescence of a building. Often a new function entails requirements which comply with existing structures more than the original function itself. Therefore, conservation cannot avoid adaptive actions, or to say better a co-evolutionary process with the environment, in which the object adapts itself to the environmental changes, and these changes are conditioned by the presence of the cultural object. That's why some years ago we proposed to think preservation in terms of safeguarding co-evolutionary potentialities of the cultural objects (Della Torre 1999).

Furthermore, far from being only a matter of material processes, conservation takes place also because of people involvement, recognition of values, a complex exchange with valorization activities (Van Balen & Vandesande 2015). Also referring to Faro convention (Therond & Trigona 2009), valorization has been defined as "the relational dimension of preservation" (Petraroia 2014). These definitions stress the need of dealing with preservation issues as a whole, never forgetting the complexity of the involved problems. This is a challenge, as actors in the heritage field are often experts and specialists (art historians, scientists, architects, engineers, archaeologists, economists etc.), who are willing to care just for their own micro-process, not for the wholeness of preservation.

3 MANAGEMENT AND DECISION MAKING

Decision making in cultural properties management suffers of some bias. On one hand, conservation and valorization are often thought as separate things, identifying valorization with tourism industry and cultural sites with theme parks. This attitude influences policies and investment strategies. On the other hand, conservation works, including major restorative works but often even maintenance, are meant as activities that go beyond the manager's skills, unrelated to values and valorization activities, and are kept out of ordinary budgets.

Cultural properties need a long-term vision, but also a tight management control, dealing with costs and revenues, as well as with energy efficiency, occupancy comfort evaluation, and conservation issues of course. Also the risks, which the property suffers, should have a periodic assessment. Depending on the findings and the risk assessment (updated following the evolution of boundary conditions) the decision will be taken to keep maintaining the building as it is, or to start the subprocess of major corrective or adaptive works.

As in all flow-charts, the decision rhombus open two separate paths. In this case, they can be thought as separate routines, that is a repetitive routine made of maintenance and monitoring activities, and a chain of programming, designing, executing exceptional works. But it will be crucial also to highlight how all these different activities should be carried out with a good consistency and coordination among them.

4 PREVENTION AND MAINTENANCE

The discussion about the specific character of maintenance implemented on built cultural heritage, and its value as a preventative factor, could lead very far. In our scheme we identify prevention as a specific field of activities, as Italian Law enforced in 2004 the separate definitions of Prevention and Maintenance. They have been developed starting from other disciplinary fields, adding other "conservation activities" to the traditional "restoration", and fostering research and field.

Following the approach enforced by Italian Law, we keep this distinction among prevention and maintenance, even if we understand maintenance as a strategy to avoid the causes of decays, deciding the maintenance interventions on the basis not just of a condition assessment, but of an evaluation on threats and hazards. So we deal separately with prevention and maintenance/monitoring, recognizing that the two activities are often interrelated.

4.1 *Prevention*

Prevention is defined in Italian framework law 42/2004, art. 29, as "the complex of the activities capable to limit the risk conditions concerning a cultural good in its context" ("il complesso delleattività idonee a limitare le situazioni di rischio connesse al bene culturale nel suo contesto").

Three topics emerge: prevention cannot be reduced to a single and simple perspective; prevention is related to risk management, thus recalling a specific technological background; risks is born by the interactions between the cultural object and the environmental and anthropic context, therefore encompassing also the threats produced by the conditions of use and valorization.

Risk analysis is usually based on formulas like threat (probability of an unwanted event) × vulnerability (magnitude of consequences, amplified because of recognized values). This approach enables to specify the activities that can be most effective in front of the various hazards, aiming at the reduction of threat or of vulnerability, or both, according to the possibility to manage the threats.

Besides the obvious references to risk management in numerous fields, a specific toolkit for prevention has been developed particularly for movable cultural heritage conserved in museums, libraries and archives (Waller 1994; Ashley-Smith 1999; Taylor 2005).

These studies produced typologies of hazards, which have been extended to immovable heritage (Waller 2003; Paolini et al. 2012). Dealing with buildings the problems are far more complex, because of two main reasons: on one hand, the surrounding environment is generally not controlled as it is in museums and libraries; on the other hand buildings have to be used, and this implies that practical issues are seldom strictly related to conservation, while they often refer to use, implying adaptive, more than curative works.

Therefore the research shows a different state of the art for the specific themes. In the reality, some risks will be dealt with at the level of the building and the functional organization of its use, some others suggest an intervention on the surroundings, if feasible. Prevention strategies will vary according to the feasibility of operating on dangerousness or vulnerability, and on the surrounding context or on the property itself.

The typical case is that of seismic hazard: for earthquakes it is possible to estimate the probability of a return with a given magnitude in a given time interval, but nobody can impede or mitigate the event. Therefore the strategy is focused on protection, that is the reduction of vulnerability, and the organization of a swift and effective mobilization. Surely a good maintenance is crucial, because in all quakes it has been observed that a bad state of repair amplifies the damages. But the reduction of vulnerability often implies interventions and changes, whose design can be more or less sensitive to conservation issues.

On this topic a great deal of research has been carried out, leading to the reasonable conclusion that interventions should be "specifically tailored for each single case, carefully considering its historical and artistic values, and designed on the basis of equilibrium considerations of every single portion/component of the structure and of "technical/economical optimization procedures" in the selection and use of the available, whatever traditional or innovative, technologies" (Modena et al. 2016).

Other natural hazards, as amplified and made more frequent by climate change, may be addressed by policies on the long run, but in the short term they require the same strategy of protection.

Among meteorological events, thunderbolts, tornados, heavy rainfall, maybe made more severe by climate change, suggest different preventive actions. Tornados cause unexpected mechanical actions diverse from static loads, and therefore such hazards require to reduce vulnerability, adding special devices if needed. As for and lightning protection and earthing systems, the technical tradition is well established also on monuments since XIX c. On the other hand, a diverse regime of rainfalls (Cassar 2005; Sabbioni, Brimblecombe & Cassar 2012; Heathcote, Fluck & Wiggins 2017) suggest to check draining systems, which sometimes prove to be under dimensioned. In other words, even if since Ruskin's times the removal of leaves from gutters is the best example of simple and direct preventive action, to clear gutters and pipes is often no longer enough, as concentrated rainfalls tend to exceed the capacity of thedrainage systemsdesigned only few years ago, which need to be enhanced more than maintained.

The risk related to flooding hazard, also powered by climate change, is mostly an exposition issue, and a matter of (literally) upstream interventions. There is nowadays a set of lessons learned about collections and archives put in places subject to flooding, while weather forecasts and hydrology enable to foresee overflowing and also to manage floods in river basins. Nevertheless coastal accelerated erosion is an issue, for which any control still seems quite difficult (Murphy, Thackray & Wilson, 2009).

Unlike seismic risk, the risk due to fire is mostly related to human behaviours. In most countries standards for fire prevention include specific norms for cultural properties, which avoid to implement ordinary measures unthinkingly, thanks to the principle of equivalent safety.

Security systems against vandalism and thefts are rapidly developing under the pressure of attacks, which target cultural heritage also because of its symbolic values. The problem has to be dealt with also as a public order question, but it impacts on the conditions for the enjoyment by people: to say better, valorization implies some risks, so that it should be promoted without passing the limits (the so called carrying capacity) suggested by conservation issues.

Other, never negligible hazards come from the human activities in the surrounding territory: pollution, traffic vibrations, shock waves due to explosions or works, subsidence due to pumping groundwater or nearby construction… In all these cases it should be possible to mitigate the causes, but it is obvious that the problem is posed at un upper level of interaction and decision.

Therefore prevention encompasses a very mixed set of activities: some of them may concern the building, and therefore pass through the decision making process of works, which exceed regular maintenance; some of them may occur in the surroundings, thus requiring an even more complex decision making at an upper level, urban or even regional; on the other hand, no prevention measure could be effective on the long run without a regular and appropriate maintenance, which keeps the devices efficient, and contributes to keep the whole system under control.

4.2 *Preparedness to risk*

In front of risks, which happen in forms not completely predictable, as earthquakes, flooding's, tornados, fires, attacks and so on, the reduction of vulnerability is not enough to avoid damages and critical situations. It is therefore useful to prepare adequate resources and tested procedures in order to intervene timely and effectively in emergency to limit the damages and to re-establish as soon as possible safety conditions.

The issue of procedures concerns both knowledge management and the clear definition of roles, skills and tasks, but the link between emergency actions and daily practices is still often missing. Rescue plans and urgent measures could be prepared exploiting the knowledge and optimizing the resources employed in maintenance

As emergency is often not limited to a single property, an established maintenance service could be very supportive in identifying the opportune human and technical resources.

Therefore, Prevention encompasses many activities, which could hardly be reduced into a system focused on maintenance, monitoring and housekeeping, as they involve a decision making that goes beyond the scale of the single property. On the other hand, in many cases maintenance seems to be the most effective strategy for prevention, or at least a necessary complement of other strategies.

4.3 *Maintenance*

The definition enforced by Italian law says that maintenance is "the complex of the activities and the interventions aimed at the control of the conditions of the cultural good and at keeping integrity, functional efficiency and identity of it and its parts" ("ilcomplesso delle attività e degli interventi destinati al controllo delle condizioni del bene culturale e al mantenimento dell'integrità, dell'efficienza funzionale e dell'identità del bene e delle sue parti").

The text is an important reference because, during its complex gestation, it broadened the understanding of maintenance from the simple keeping integrity and efficiency to condition assessment and safeguarding of identity (Della Torre & Gasparoli 2007). The first point has relevant consequences on the quality of maintenance activities, as the request of performing an assessment entails high skills and experience even for the simplest operations; furthermore, also monitoring activities and inspections can be classified as maintenance.

As for keeping identity, this distinguishes the maintenance of cultural heritage from that implemented on ordinary objects, as the requirements of integrity and efficiency get conditioned by a third issue, related to authenticity concept, which implies the substitutive practices typical of scheduled predictive maintenance in the industrial sector.

For maintenance sake, it is crucial to define the standard levels of integrity and performances the building has to comply with. Activities may aim just at avoiding the insurgence and propagation of damages, or at remaking a high quality image, which probably would end into the most critical questions of restoration, and into heavier and inopportune alterations of historic fabric.

The evaluation of preventive efficacy is not always easy, due to the complexity of causal chains, but this criterion could really guide among the alternatives of maintenance works. If it should be clear that maintenance for cultural heritage should always be based on assessment, the step from the plain condition assessment to a diagnosis focused on processes, threats and hazards is crucial, as the one that moves maintenance practices towards the cognitive field of preventive conservation. The difference among maintenance and prevention is therefore to be found in the things to do: when detected risks require major interventions, or actions in the context, it is Prevention; when actions keep in the limits of ordinary practices, it is Maintenance.

Planning consistent and coordinated inspection and maintenance activities stays in a frame of management choices, spanning from the definition of the carrying capacity of the sites and the single rooms, to the translation of this capacity into thresholds of measurable quantities, in function of recognized values, risks and valorization policies. Management plans and systems should always be prepared when restoring buildings, taking into account the future uses in sight of assessing project choices on the long run. It is worthy to notice that most of the assessments required by management control are the same that would be carried out in the frame of a conservation plan.

A further issue concerns the availability of required skills, in house or at least on the local market. In the past experiences, pure maintenance activities did not prove to be profitable enough to trigger the market, but they surely could yield externalities, which once acknowledged and valorized would be able to give an important contribution to the feasibility of positive innovation (Della Torre 2015).

These are both of agglomeration externalities and Jacobs externalities, and the prevailing of one or the other character will depend on the capability to see the complexity and the opportunities of the site (Caragliu, de Groot & De Dominicis 2016).

5 PROGRAM, MAJOR WORKS AND OVERALL CONSISTENCY

To model the interventions splitting them into three phases, i.e. program, design and execution, helps to emphasize the importance of consistency and planning. Interventions should not be simply executed, or designed and executed, but they should be carefully prepared in order to guarantee a set of requirements and benefits on the long run.

The phase of program deals with all the criteria to be implemented, e.g. for theselection of professionals and companies, and is usually focused at least on financial feasibility and costs control. Also this starting point can be exploited, if there is awareness about the potential benefits related to different funding strategies. Funds are not equally productive: to get the necessary money is the basic condition to get going, but some experiences prove that more sophisticated business plans, based on strategic alliances, not only point out how to gather money, but produce also long-term cooperation strategies, which make the projects far more sustainable (CHcfE Consortium 2015).

Furthermore, in the program the conditions for higher quality can be defined, as well as a better organization for the aftercare can be established. In other words, if there had been a lack of prevention, which made major curative works unavoidable, these works could be programmed in order to produce conditions for a better conservation in the future. Here program will be discussed as a tool for quality, for conservation planning, for sharing a long term vision.

5.1 *Program as a tool for quality*

It is possible to speak of "quality" when the discourse is referred to a set of requirements set in advance: the construction briefingis a core component of the program (Barrett & Stanley 1999), in which all the requirements are clearly written and agreed by the stakeholders in the construction process. Also for refurbishment or restoration works it is useful to clarify as soon as possible the targets of the intervention, and the high quality label will mean that the output complies with the foreseen requirements.

Also the criteria can be defined, which will be followed for thedesign and the execution of the works. Here again, quality may not be a generic acknowledgement, and the adoption of best methods and tools should comply with a clearly stated prevision. The use itself of advanced methods is usually expensive, then should be optimized. The best survey, and the best diagnostic campaign produce data and knowledge, but this doesn't produce a "quality project" without the proved usefulness of such data.

The program phase is crucial to establish a shared awareness about what quality means and what is required to set up a quality process. This is often a problem of resources, also in terms of cash-flows. Precision surveys and diagnostics cost, and above all their output is necessary in the very beginning of the process, when money is often still to be found. So the problem is twofold: to convince decision makers that these expenses are productive, and also that the first investments should be more expensive than customary because of the advantages to be seen in the future. It should be noticed that in these kind of activities owners tend to waste resources because of failures in knowledge management: if the reports from previous works were available for an easy information retrieval, diagnostic campaigns could be more focused on selected targets, and then less expensive. This is always a knot to be dealt with while developing the program phase.

5.2 *Program and conservation planning*

Even if targeted on interventions, the program has to implement a broader vision, taking into account the uses, the facility management system, the overall sustainability. Therefore, if a continuous care system is not available on the spot, in the program phase it is possible to deal with this issue, choosing the kind of management plan required as a part of the project.

The implementation of an information system could be decided as well at this level, and it is the only trustable level because the first requirement for such a system should be to connect in an interoperable way all the phases and all the players of the conservation and valorization process: and only in this phase of planning they are all at hand.

In other words, the program phase is the one dealing with present and future budgets. The check on the convenience of setting up a preventive system, implementing digital tools and investing in preparedness to risk could lead to a choice at the same level of the other choices related to enhancing performances or managing risks.

5.3 *Systemic and long-term vision*

As the phase that prepares the project, the program is based on contents, which are similar to the ones typical of business plans: recognition of values, valorization and business strategies (people involvement, target groups, alliances, etc.), funding strategies, evaluation methods for market and non-market values, often advocacy for preservation.

In this responsibility for long-term sustainability is the link among the major intervention and the management aware ofthe preventive approach.

The program should take care that resources don't go wasted in interventions, which have not be thought in the frame of a careful, compatible and sustainable use, and encompasses the tools to turn the intervention into an opportunity for a durable change.

As a perfect complement of maintenance and monitoring techniques, program can be the tool to ensure financial support and strategic alliances. Experiences like the Halland model in Sweden and Distretti culturali in Italy (both referred to in CFcfE 2015) prove that the more sustainable projects build networks before (upstream), that is in the program phase, enabling the projects to be cross-sectoral and to exploit benefits, which otherwise would be pure externalities. The Distretti culturali projects has been indeed a huge exercise of business plans and construction briefing, as for the first time actions were funded on the basis of complete programs, and no longer on the basis only of restoration projects. Moving the emphasis on the completeness of the program, the decision maker got better guarantees on the feasibility and sustainability of the proposed action.

6 INSPECTION, KNOWLEDGE MANAGEMENT AND DECISION MAKING

Each intervention on built cultural heritage should produce feedback in terms of a scientific report and instructions for maintenance, to be implemented in a general plan. An Information system could file the knowledge produced as designing the intervention, curriculum of the building and each single component, the works done and the controls and maintenance interventions suggested for the future.

Maintenance itself has to be thought non as a simple activity not to be burdened with further requirements, but as one of the activities that produce conservation, if they are coherent, coordinate and planned. In this logical frame resources for maintenance shall be scheduled and coordinated with the other activities, and the information output shall go into a systemic information management, to gest consistent with the other phases. This holds even more if maintenanceis correctly understood as a sequence of controls and inspections, ending up in minimal interventions just required to avoid or to stop unwanted processes, thus producing above all information (Figure 1).

Therefore, as for buildings and sites the number and complexity of data and their interactions is definitely huge, various methods have been tested, mostly adopting relational databases. The frontier of this tendency id informative digitalization, encompassing surveying, design and management through interoperable techniques. Such a technological transfer from the construction industry to preservation domain is already in action.

Research on the so called Historic-BIM had a rapid development in the last few years (Volk, Stengel, Schultmann, 2014; Logothetis, Delinasiou, Stylianidis, 2015; Megahed, 2015), though sometimes focusing more on 3D models than on the interoperability and the uniqueness of any datum in a Common data Environment.

It is noticeable that on the crucial knot of the definition of LODs (Level of Development, or Level of Detail; sometimes splitted in Level of Geometry and Level of Information) for processes concerning historic buildings, the first reference norms issued are consolidating a processual vision, according to which information is gathered from previous intervention, or is built from scratch as soon as possible at a very high level of accuracy, refusing the incremental logic typical of new construction.

Reasoning on this turnover of the development flow, it is obvious to end up on an issue already underlined by BIM research, that is the distinction among the concept a (unique) interoperable model and the concept of a central archive (usually referred to as Common Data Environment), a kind of platform enabling the exchange among several "domain specific" BIMs.

These specific models (e.g. BIM for the architectural design, for structural analysis, for construction and site management, for energy simulations, for cost analysis, facility management, and so on) are perfectly identified as BIMs, surely more performing and manageable than the enormous single universal model, which should also serve as the exchange platform. Each domain specific model

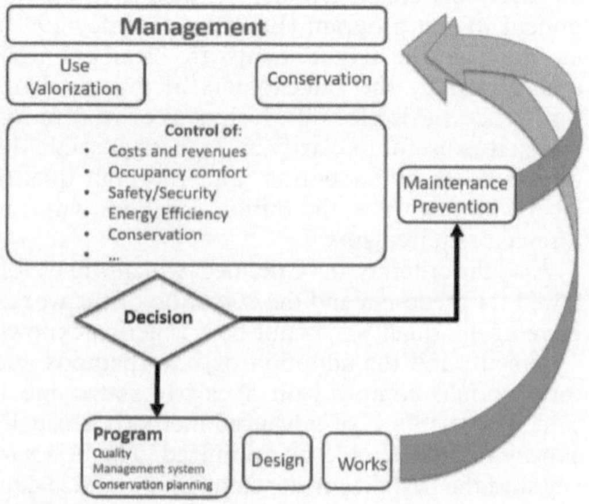

Figure 1. Layout of the process.

should download form the general database only the useful data, and then upload its own output to make it available to other activities, always through the CDE, which obviously developed its meaning, from its first definition in British PAS, in the direction of becoming a real relational database.This could be a direction for next research, focusing on exchange formats (Oreni et al. 2017) and on the dynamics among CDE and single BIMs, working as well on costs, times and skills required to make this evolution happen.

This vision of several BIMs referred to one CDE has been theorized also as the way to ensure the highest interoperability (Laakso & Kiviniemi 2012). But ii is also the way to support a crucial change in heritage sector, because the required anticipation of accurate investigations, as well as the availability of tools enabling a continuous monitoring, will give the whole process consistency, coordination and concrete opportunity to make activities planned.

7 CONCLUSIONS

The contribution tried to analyze the whole conservation process, in order to understand the relationships among the activities, which are usually identified with "preventive conservation" and the more traditional ones, under the hypothesis that all the activities contribute to conservation if they are correctly planned.

In this vision, the alternative between preventive activities and curative actions is not ideological, but the output of practical decision based on management control and risk assessment.

This means that historic buildings require a management system, which could be designed as a conservation plan including facility management issues. It is worthwhile notice that the management control requires specific items of information, which often are missing, unless a correct maintenance system, based on regular and structure inspections, is implemented. To enable a correct and informed decision making, it should be complemented with customary Facility Management controls, but above all with risk assessment. Given the diversity of the hazards and their different scale, risk assessment require a vision which goes far beyond the borders of the single property.

Controls and risk assessment can produce findings, which require corrective or adaptive works. In this cases, an analysis model split into three blocks allowed to emphasize the role of the program phase, as the one which has the broadest and more systemic vision, and thus can deal with the issue of consistency, interoperability and long-term planning, turning the major interventions into opportunities for changes in the direction of implementing more efficient tools for preventive conservation.

REFERENCES

Ashley-Smith, J. (1999). *Risk assessment for object conservation*. Oxford, Boston: Butterworth-Heinemann.

Barrett, P. & Stanley, C. (1999). *Better Construction Briefing*, Blackwell Science: Oxford, 1999.

Bellini, A. (1998). La pura contemplazione non appartiene all'architettura, *Tema*, 1/1998: 2–3.

Caragliu, A., de Groot, H.L.F. & De Dominicis, L. (2016). Both Marshall and Jacobs were Right!, *Economic Geography*, 92(1): 87–111.

Cassar, M. (2005). *Climate Change and the Historic Environment*. London: UCL.

CHCfE Consortium. (2015). *Cultural Heritage counts for Europe*. Krakow: International Cultural Centre.

Della Torre, S. & Gasparoli, P. (2007). La definizione di manutenzione nel Codice dei Beni Culturali: un'analisi del testo e delle sue implicazioni. Riferimenti e confronti con le attività manutentive sul costruito diffuso", in V. Fiore (ed.), *La cultura della manutenzione nel progetto edilizio e urbano*: 160–163. Siracusa: LetteraVentidue.

Della Torre, S. (1999). "Manutenzione" o "Conservazione"? La sfida del passaggio dall'equilibrio al divenire. In G. Biscontin & G. Driussi (eds.), *Ripensare alla manutenzione. Ricerche, progettazione, materiali, tecniche per la cura del costruito:* 71–80. Venice: Arcadia ricerche.

Della Torre, S. (2015). Shaping Tools for Built Heritage Conservation: from Architectural Design to Program and Management. Learning from 'Distretti culturali', in K. Van Balen & A. Vandesande (eds), *Community Involvement in Heritage*: pp. 93–102. Antwerp—Apeldoorn: Garant.

Heathcote, J., Fluck, H. & M. Wiggins. (2017). 'Predicting and Adapting to Climate Change: Challenges for the Historic Environment', *The Historic Environment: Policy & Practice*, 8:2, 89–100.

Laakso M. & Kiviniemi A. (2012). The IFC standard—a review of history, development, and standardization, *Journal of Information Technology in Construction (ITcon)*, Vol. 17: 134–161.

Logothetis, S., Delinasiou, A. & Stylianidis, E. (2015). Building Information Modelling for Cultural Heritage: a Review, ISPRS Annals of the Photogrammetry, Remote Sensing and Spatial Information Sciences, Volume II-5/W3, 2015 25th International CIPA Symposium 2015, 31 August–04 September 2015. Taipei, Taiwan.

Megahed, N.A. (2015). Towards a Theoretical Framework for HBIM Approach in Historic Preservation and Management, *International Journal of Architectural Research*, Volume 9—Issue 3: 130–147.

Modena, et al. (2016). Modena, C., da Port, oF., Valluzzi, M.R., Carapezza Guttuso, F., Iannelli, P. & Rubino, C. Sustainable approaches to the assessment and mitigation of seismic risk and of the effects of earthquake induced damages to historic urban centers.

In K. Van Balen & E. Verstrynge (eds), *Structural Analysis of Historical Constructions—Anamnesis, diagnosis, therapy, controls*: 35–43. London: Taylor & Francis Group.

Murphy, P., Thackray D. & Ed Wilson. (2009). Coastal Heritage and Climate Change in England: Assessing Threats and Priorities, *Conservation and Management of Archaeological Sites*, Vol. 11: 9–15.

Oreni, D., Brumana, R., Della Torre, S. & F. Banfi. (2017). Survey, HBIM and Conservation Plan of a monumental Building damaged by Eratquake. The International Archives of the Photogrammetry, Remote Sensing and Spatial Information Sciences, Volume XLII-5/W1, 2017. GEOMATICS & RESTORATION—Conservation of Cultural Heritage in the Digital Era, 22–24 May 2017, Florence, Italy.

Paolini et al. (2012). Paolini, A. Vafadari, A., Cesaro, G., Santana Quintero, M., Van Balen, K., Vileikis, O., Fakhoury, L., *Risk Management at Heritage Sites: a Case Study of Petra World Heritage Site*. Paris: UNESCO.

Petraroia, P. (2014). La valorizzazione come dimensione relazionale della tutela, in G. Negri-Clementi & S. Stabile (eds.), *Il Diritto dell'arte. 3. La protezione del patrimonio artistico*: 41–49. Milan: Skira.

Sabbioni, C., Brimblecombe, P. & Cassar, M. (eds.) (2012). The Atlas of Climate Change Impact on European Cultural Heritage: Scientific analysis and management strategies, London: Anthem Press.

Taylor, J. (2005). An integrated approach to risk assessment and condition surveys. *Journal of the American Institute for Conservation*, 44: 127–141.

Therond, D. & Trigona, A. (eds) (2009). *Heritage and Beyond*. Brussels: Council of Europe Publishing.

Van Balen, K. & Vandesande, A. (eds) (2015). *Community Involvement in Heritage*. Antwerp—Apeldoorn: Garant.

Van Balen, K. & Vandesande, A. (eds.) (2013). *Reflections on Preventive Conservation, Maintenance and Monitoring of Monuments and Sites by the PRECOM³OS UNESCO Chair.* Leuven: ACCO.

Vandesande, A., Moioli, R. & Van Balen, K. (2014). Costing the Built Environment: Towards a Policy of Strong Sustainable Development. In G. Biscontin & G. Driussi (eds.), *Quale Sostenibilita' per il restauro?*: 457–467. Venice: Arcadia Ricerche.

Volk, R., Stengel, J. & Schultmann, F. (2014). Building Information Modeling (BIM) for existing buildings—Literature review and future needs, *Automation in Construction*, 38, 109–127.

Waller, R. (1994). Conservation risk assessment: A strategy for managing resources for preventive conservation. In *International Institute for Conservation, Preprints of the Ottawa Congress*, September 1994: 12–16. London: International Institute for Conservation.

Waller, R. (2003). *Cultural property risk analysis model*, Göteborg studies in conservation, 13. Göteborg: Acta Universitatis Gothoburgensis.

Innovative Built Heritage Models – Van Balen & Vandesande (Eds)
© 2018 Taylor & Francis Group, London, ISBN 978-1-138-49861-7

Planning the re-use: Sustainability through resilience evaluation

M. Morandotti
Department of Civil Engineering and Architecture (DICAr), University of Pavia, Pavia, Italy

ABSTRACT: The paper presents first methodological achievement of an ongoing research aimed at the development of an evaluation tool aimed to support sustainable management of historical building through resilience evaluation. The "Resilience Evaluation Method" (R.E.M.) may be applied in order to evaluate the average level of use compatibility between a settled function and an existing building. A sustainable refurbishment strategy shall be focused on an adequate preliminary evaluation of the possible impact of the new function or the impact of the spatial and technological changes due to the necessity to adequate the existing building to the users' needs. Final aim is to define both the expected impact of the new function on the existing building, and the maximum allowed rate of transformation of the existing building.

1 INTRODUCTION

The paper deals with the general topic of cultural heritage preservation and valorisation. It focuses on sustainable restoration, and in a more specific way, takes into account resilience assessment and evaluation of an existing building in the perspective to guarantee a less intrusive, more responsible and more sustainable restoration project.

Pre-determination of the impact level of new functions on the existing building is crucial in the perspective of a sustainable approach to the conservation.

Recent debate on social, ecological and sustainable urban development increasingly includes resilience amongst its fundamental categories. The attitude of a territory, a city, or a system to adapt and to respond positively to the changes and demands of the context, or "the capacity to lead to a continued existence by incorporating change" (Folke et al. 2003)—may be recognized as one of the primary values in a sustainable evolutionary perspective.

According to the general theory regarding the resilience of complex socio-ecological systems (Folke et al. 2010) modifying pressures, the so-called drivers of change, generate perturbations in response to which the system demonstrates its higher or lower resilience, given precisely by its ability to cope with the change that has been introduced while maintaining its own constitutive identity, without setting out on transformative trajectories that lead to less desirable conditions.

In the context of urban and urban-sociological studies, resilience is often seen as a privileged interpretive key in the development of a vision that is suitable for interpreting the systemic changes currently taking place, and for identifying possible adaptation strategies with respect to, for example, climate change, the reduction of natural resources, and extensive transcontinental migration phenomena. It is in fact a key tool for understanding the levels of modifying stress with which urban socio-ecological systems can cope without evolving towards less desirable forms, both environmentally and socially.

Strictly within the field of civil engineering, the subject of increased resilience in immovable cultural heritage is mainly developed in relation to anthropogenic and natural risk management procedures (Jigyasu et al. 2012) pertaining to calamitous events. A specific declination is the application of the concept of resilience (in its mechanical meaning) to the issue of reducing the seismic vulnerability of existing structures, with particular reference to historical ones.

Moreover, the preservation and protection of historical heritage is an effective strategy for increase the resilience of territories, due to the fact that historical heritage is deeply involved in defining cultural identity of each community. Solid identity self-consciousness, combined with flexible and decentralized aggregative structures, large levels of functional self-sufficiency and partial redundancies of infrastructures and services, constitute the central elements in each of the medium and long term strategies aimed at increasing territorial resilience.

In the following, resilience will be defined as the tolerable transformation that a generic existing building can undergo unless the impact on its constitutive systems (material, typological, distributive, constructive) generates undesirable effects on its semantic coherence, if not its physical existence. This transformation ensures that the building system can adapt and accommodate changes arising from new uses, or from performance and regulatory updates to previous uses while maintaining a recognizable identity and a functionality level appropriate to new uses and respectful of the historic value of the asset.

The identification of the macro intervention categories, defined in terms of performance, is based on a set of parameters that provide a measure of the expected deviation from the acceptability threshold. In the general theory of resilience, these indicators appear as control variables, or sentinel values. Their definition is of great relevance regarding the method's efficacy and operability, especially in relation to the possibility to identify a value limit for each variable that constitutes the "acceptability threshold of transformations" (Fabbricatti 2013). In this perspective, the concept of resilience is intrinsically linked to the theoretical and methodological approach of the sustainable re-use of historic building heritage and, therefore, with the valorisation of cultural heritage.

The transformative pressure upon existing historical assets is nowadays generated by at least three distinct factors: i) the push towards the re-use of assets through the definition of a new set of functions; ii) the push exerted by binding legal provisions (for example in the field of safety and fire prevention, the reduction of seismic vulnerability, accessibility) whose requirements may result from new functions to be relocated in existing spaces; iii) the push exerted by the need to improve the energy behavior of the building. This transformative pressure is commonly translated through design action, in interventions of performance improvement of a functional, distributive, technological, structural, or plant engineering nature. These could lead to the actual fulfilment of users' needs, but may also generate significant impacts on the existing historic object.

It is today possible to deal with continuous updating of immovable cultural heritage, in a technological or typological perspective. In particular, upgrades to building performance concerning health (internal comfort), usability (the insertion of new mechanized vertical connections), and security may be included in the first category. Here also belong not only technological updates related to fire safety, but also structural upgrades aiming at the reduction of the seismic vulnerability of existing structures. Less easily measurable, however, are actions attributable to the typological updating of existing assets, such as those of controlled densification for improving the exploitation of residual spaces or, more generally, the morphological reconfiguration of spaces, otherwise known as adaptive re-use.

2 METHOD AND TOOLS

The method described (Fig. 1) is focused to assess the overall sustainability of intervention in relation to specific resilience thresholds of the system that

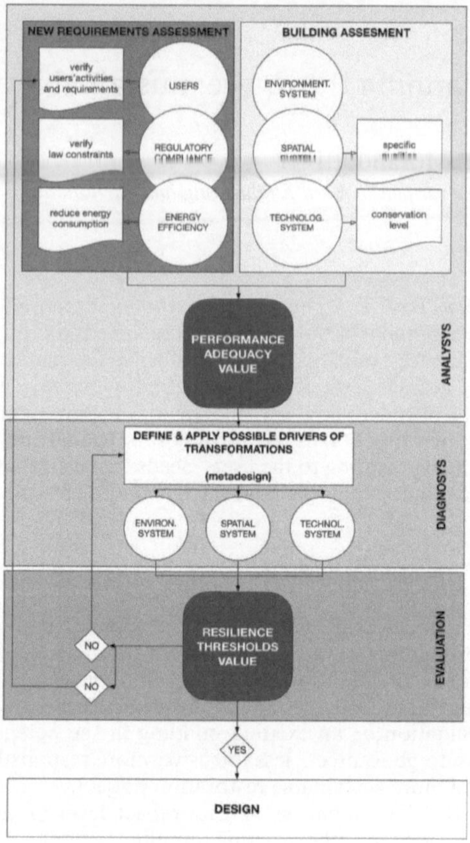

Figure 1. General flowchart of the method described.

are determined by the balance of requests for conservation, the satisfaction of users' needs, and the managerial sustainability of the process. It may help decision makers and stakeholders involved to avoid forced placement of inadequate functions in existing buildings. Otherwise a design solution, albeit coherent with new needs and with current regulations, may generate drastic compromises of material, typological, or pre-existing technological integrity, and therefore become, ultimately, un-sustainable.

The method proposed is developed according to the sequence of analysis/diagnosis/evaluation. The general scheme of the process developed includes two synthetic indicators of behavior or performance. The first one has been defined Performance Adequacy Value (PAV), and measures the performance analysis evaluation of the existing building. The second one has been defined Resilience Threshold Value (RTV), and measures the impact of transformation design solutions on the existing building.

Two interacting systems are analysed: the perturbative one, consisting of the heterogeneous set of different transformative pushes in terms of functions, activities and requirements generated by stakeholders, end-users and the community, and the perturbed one, consisting of the building system being investigated. This phase adopts classic methods for acquiring knowledge about the historic object in terms of surveys, mapping and diagnostics, all examined at appropriate levels in order to develop a preliminary evaluation. The analysis will be deepened in the following stages of the design process, but in the meanwhile it will provide a solid interpretative framework of the main building subsystems and the overall conservation status of the object. Starting from a morphological, structural, technological, material and strictly conservative reading of the constraints peculiar to the historic building, a parallel analysis is developed, based on the criteria of the meta-design and performance approach, according to the well-known trinomial needs-requisites-performance. This comes from the set of needs related to planned destinations and the generated environmental and technological requirements. The proposed approach envisages the development of a critical investigation of the building concerning both spatial survey, and major technological systems.

Spatial investigation deals with dimensional aspects of the building, taking into account available areas and connecting systems, both horizontal and vertical. Distributive systems' analysis will assess critical aspects dealing with accessibility and fire security as both may generate either severe limitations in use, or huge building's transformations.

Spatial and material analysis must be integrated by a similar overview of the environmental and technological systems before intervention. This effort is due to define the set of residual environmental benefits provided by the building, both concerning the performance level of the structural envelope, and effectiveness of orientation concerning natural lighting in confined spaces.

Analyses of users' needs, divided into environmental/spatial systems and technological systems, are aimed not only at identifying a specific response but also at placing this response against the respective system of previously determined limitations.

To summarize the results of the analysis on the existing buildings, has been defined the Performance Adequacy Value (PAV). It is calculated applying specific parameters which synthetize the complex system of spatial requirements into four indicators: i) usability, intended as both dimensional and distributive adequacy; ii) comfort, both thermal and luminous; iii) safety, both as seismic vulnerability and fire security; iv) accessibility.

It has been defined a specific evaluation chart, (Fig. 2) which provides an evaluation score in a range between 0 (absent) and 3 (good), according to a correspondent performance range. The unweight average valuation of each element determines the level of performance adequacy of the building. The greater the indicator value, the lower the transformation pressure, and consequently the transformative stress.

Usability is evaluated taking into account dimensional and distributive adequacy.

Dimensional adequacy is defined as available and necessary area ratio. Necessary area in determined, for each specific function which his intended to be settled in the existing building, referring to general dimensioning standards, either defined by compulsory requirements, or determined according to literature parameters. The final value of the parameter may vary, even sensibly, according to different functional reuse hypothesis. Defined performance ranges assign low score not only to under dimensioned spaces, due to possible inefficiency generated by crowded work spaces, but also to over dimensioned ones, as the latter scenario may represent a lack of sustainability in terms of inadequate space management optimization.

Distributive adequacy refers to both horizontal and vertical connective spaces. The former is defined by the ratio between the area connected by corridors and the total area of the building. It is a useful parameter to evaluate the presence and the effectiveness of the distributing system.

Accordingly to the specific function required, it could become a relevant evaluation parameter. For example, for offices or university administrative departments, a poor value of this indicator means that the project would probably solve huge distributive problems, eventually by means of relevant spatial and material transformations

Adequacy of vertical connection systems is evaluated by the distance between existing stairs. Few stairs, very distant each other will probably generate relevant problems, not only from a functional point of view, but also in the perspective of safety, security and accessibility compulsory requirements.

Comfort is considered both as thermal and visual. Thermal comfort is defined as the ratio between required and actual transmittance of the existing facades. Although in an existing building this specific value may vary sensibly from point to point, as the proposed one is a simplified, pre-design assessment, it seems acceptable to assume the actual transmittance value as the value calculated in one section, estimated as the most common one in the building envelope. If the assessment phase includes several material surveys of the building envelope (depth, stratigraphy, materials) than the actual transmittance value may be assumed as the average value among the calculated ones. Although in case of historical buildings it is not compulsory to reach a determined level of energy efficiency, a design strategy oriented to energy sustainability will anyway try to reduce the carbon footprint of

requirement			parameter	performance range			assigned score
USABILITY	dimensional adequacy		avalilable surface/necessary surface	≤ 0,2	or	>1,2	0
				>0,2 ≤0,5	or	>1,1 ≤120	1
				>0,5 ≤0,8	or	>1 ≤1,1	2
				>0,8 ≤1			3
	distributive adequacy	corridors	area connected by corridors/building area (*)	≤ 0,2			0
				>0,2 ≤0,5			1
				>0,5 ≤0,8			2
				>0,8 ≤1			3
		stairs	distance between existing stairs	no stairs			0
				>50 m			1
				>30m ≤50m			2
				≤30 m			3
COMFORT	thermal		required transmittance/actual transmittance (*)	≤ 0,2			0
				>0,2 ≤0,5			1
				>0,5 ≤0,8			2
				>0,8 ≤1			3
	visual		room's window area/room's area ≥ 1/8 (*)	% rooms verified		≤ 0,2	0
				% rooms verified		>0,2 ≤0,5	1
				% rooms verified		>0,5 ≤0,8	2
				% rooms verified		>0,8 ≤1	3
SAFETY	stuctural	seismic vulnerability	D.M. n.65/7.3.2017, All.A (*)	F,G			0
				D, E			1
				B, C			2
				A+, A			3
	fire	risk level	guidelines 1/2016; DM 3/8/2015 (*)	high			1
				medium			2
				low			3
ACCESSIBILITY			building's accessible area/building's area	≤ 0,2			0
				>0,2 ≤0,5			1
				>0,5 ≤0,8			2
				>0,8 ≤1			3

(*) parameter defined according to specific Italian regulations. Application to different Countries may differ if local compulsory rules are provided

Figure 2. Evaluation chart for score assignment of the building performance.

the building, starting from its energy consumption. The lower the performance lack to be filled, the lower the possible impact on the building.

Visual comfort is determined as the ratio between the window area of each room, and the area of the room itself. This ratio is considered satisfactory (according to Italian standards) when its value is higher than 1/8. The global assessment of this requirement on the whole building is defined ad the percentage of rooms which satisfy the overmentioned ratio.

Safety deals with structural and fire vulnerability. Structural safety is defined as the level of seismic vulnerability, calculated as a recent Italian law (D.M. n.65, 7/3/2017 Annex. A). It is provided a very simplified method to assess the level of seismic vulnerability of an existing building, which can be related to specific dangerousness of the site in order to obtain an assessment of the risk class of the building itself.

Fire safety is defined by means of the Italian regulation concerning fire risk level (D.M. 3/8/2015; guidelines 1/2016), and classified in three level, as low, medium, or high. This parameter may be relevant in the analysis phase as fire security regulations are very strict, and may have a huge impact on the existing building in the design phase.

Accessibility is defined as the ratio between accessible areas of the building and the whole area. This figure may describe, even in a very synthetic and simplified way, the actual accessibility of the building before the design transformation. The lower this parameter, the higher the possible impact of the design itself on the building structure.

Some of the over described parameters (in particular comfort and safety) are defined according to specific Italian regulations, as the method has been settled to be firstly applied to Italian scenario. Although specific application to different Countries may differ if local compulsory rules are provided, the whole methodological approach will still be applicable.

Further and more detailed assessment may be developed during the design phase, but in this preliminary evaluation, this figure represents a synthetic yet effective instrument to quickly estimate the residual performance level of the existing building before the transformation.

The diagnostic phase is aimed at identifying the drivers involved in the transformation process. They are interpreted as meta-design solutions which—if carried out—would generate alterations to physical identity elements. Inherent in this approach is the sense of an early impact evaluation tool, coherent with real design process.

The ultimate aim is the identification of meta-design solutions whose overall impact, in terms of the expected transformation, is more compatible with the resilience threshold of the building. By the application of this method, specific design actions may be evaluated and compared since a very early stage, in order to support solutions which impact on the building is lower, and do not irreversibly compromise its material and documental consistency.

Operationally, the proposed method foresees that meta-design solutions defined in the previous diagnostic phase are evaluated on the base of eight possible design action on the building. Four of these, if applied, will produce a negative impact on the structure and four a positive one.

The eight control variables identified (the first four having negative impacts; the second four positive) are the following: A) loss or removal of material and the degree of loss; B) structural alteration of the building plan, and its degree of reversibility; C) spatial alteration of the building layout, and its degree of reversibility; D) loss or weakening of the structure's identity; E) increased safety; F) reduction in the environmental footprint; G) increased accessibility and coherence to design for all principles; H) increase of future adaptability and its degree of reversibility. Through a yes/no answer concerning the applicability of each action to specific design strategy under evaluation, it is possible to draft a simplified decision tree (Fig. 3). Once completed, it shows at a glance if the supposed transformation patterns are expected to have a positive or negative impact on the building. The complexity of the intervention project on the structure, however, means that the reduction to a simple 'yes/no' answer is not completely suitable. What is required are multi-criteria analysis procedures that take into account the reciprocal relationships between each control parameter and the others.

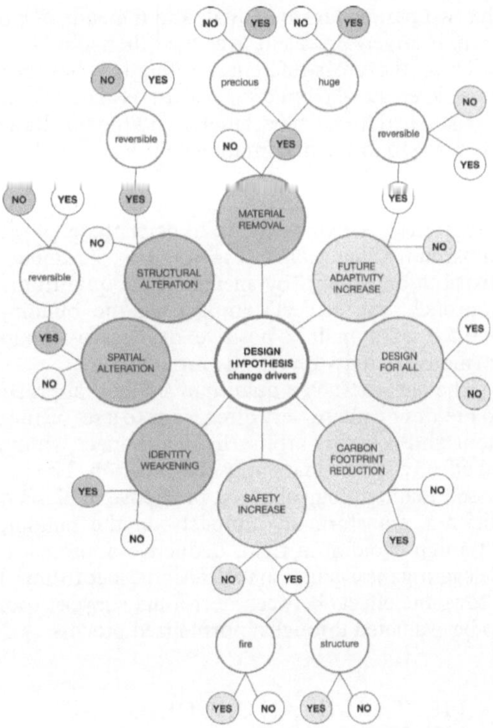

Figure 3. Decision tree, showing the eight possible design action. Each of them may be applied or not to the specific context, and therefore may generate or not related positive or negative impacts.

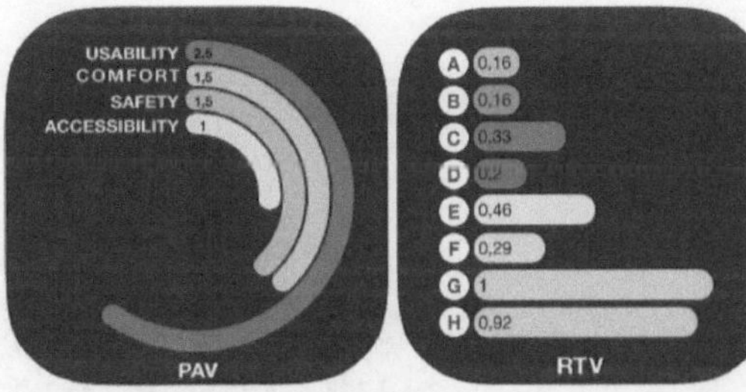

Figure 4. Graphic representation of the Performance Adequacy Value (PAV) and Resilience Threshold Value (RTV).

Therefore the model requires the compilation of a cross-reference evaluation chart once the eight-parameter decision tree analysis has been completed.

The chart is based on pairwise comparison methodology. In this way, each control parameter, if present on the previous decision tree analysis, is compared directly with the others, indicating the prevailing parameter for each comparison and its level of prevalence on a scale of 1–6 where 1 means the two parameters are equal, and 6 means one of them is largely prevalent over the other.

Once the cross-reference evaluation has been completed, an algebraic sum of the assigned values is calculated for each parameter and the results are normalized to the maximum value.

Through this process, the Resilience Threshold Value (RTV) is obtained, where each of the eight parameters is expressed as a percentage of the maximum value scored (Fig. 4). The parameters' distribution shows, by means of a quantitative approach, the project's impact on the building, defined as a mainly positive or negative transformative activity on an existing structure.

The lack of such a quantitative-based approach to pre-design phase, very often leads to a set of functions and activities arbitrarily determined, without an effective preliminary impact evaluation. The proposed methodology allows a predictive analysis of different transformative impacts on the building, although avoiding a rigid, deductive approach or delegating the evaluation to decisional algorithms. It allows the effects between decisional support tools to be evaluated through a normalized process.

3 FIRST TRIAL: AN ONGOING EXPERIENCE

The proposed method has been until now applied to historical university heritage. University building are a particularly favourable experimental fabric and are well suited to the experimental verification of investigative methodology and predictive modelling. The pressure leading to the transformation and to the functional and spatial updating of university assets can be generated by two different drivers, one internal and the other external, very specific for the university function, although the following considerations may be extended to a wider perspective.

Historical universities are often settled in monumental structures, which enrich their identity, but meanwhile generate severe constraints to their transformation. Continuous requests to upgrade have been supplemented by the need to improve management efficiency. Finally, the progressive extension of activities external to teaching and research but are being increasingly offered by universities should not be forgotten. In this regard, let us consider the varied scenario of the so called third mission, which entails a new and challenging set of requirements.

Universities are expected to open themselves to external economic, productive and social stakeholders. For these reasons historical university heritage seems a good candidate to test the over described method.

First case studies have been selected among the built heritage of the University of Pavia, which is a small-medium sized city (circa 70,000 inhabitants) and hosts an ancient university, settled in 1361,

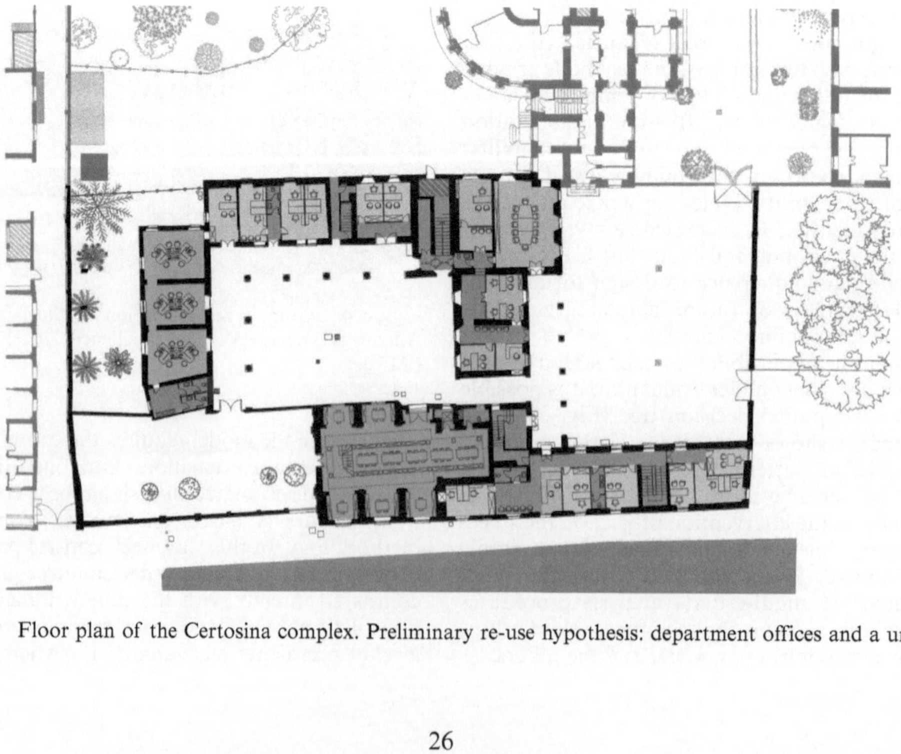

Figure 5. Floor plan of the Certosina complex. Preliminary re-use hypothesis: department offices and a university library.

requirement	parameter			performance range	assigned score		value
USABILITY	dimensional adequacy		avalilable surface/necessary surface	>0,8 ≤1	3		2,5
	distributive adequacy	corridors	area connected by corridors/building area	>0,5 ≤0,8	2	2	
		stairs	distance between existing stairs	[illegible]	[illegible]		
COMFORT	thermal		actual transmittance/required transmittance	>0,2 ≤0,5	1		1,5
	visual		% room's window area/room's area ≥ 1/8	>0,5 ≤0,8	2		
SAFETY	stuctural	seismic vulnerability	D.M. n.65/7.3.2017, All.A	D, E	1		1,5
	fire	risk level	guidelines 1/2016; DM 3/8/2015	medium	2		
ACCESSIBILITY			building's accessible area/building's area	>0,2 ≤0,5	1		1

Figure 6. Evaluation chart for score assignment of the building performance, completed after the building assessment.

that is nowadays quite notable in size (22,000 students, 50 buildings, 260,000 square meters buildings gross area). Today, the University of Pavia is committed to a series of building interventions. The majority of these are requalification projects of its historic building heritage and only in part concerning new structures' edification. A first test has been carried out applying the proposed method in the perspective of a possible future reuse of an historical building located in the city center, in order to evaluate the general sustainability of that choice in a resilience oriented perspective (Fig. 5).

In particular, it has been supposed to transfer in this actually under-used building a part of the department of Humanistic Studies.

The evaluation process has been developed, verifying a satisfactory usability performance of the building, in terms of available spaces and distributing system (Fig. 6). Comfort indicators are satisfactory too, mostly concerning visual comfort, rather than hygrothermal one. Safety level are on par with expectations, as the seismic vulnerability is high, while the Figure 7. Cross-reference evaluation chart completed for the case study fire risk is average. Accessibility on the contrary, is critical, due to the huge amount of existing barriers. The specific PAV evaluation chart shows these results in more detailed way.

Early stage design has been developed, in order to satisfy new users' needs. Resilience evaluation developed on this first hypothesis gives positive results concerning possible impact of supposed design activities on the building itself. The design strategies evaluation shows that identity weakening actions are missing, even if huge material removal and spatial alteration are foreseen.

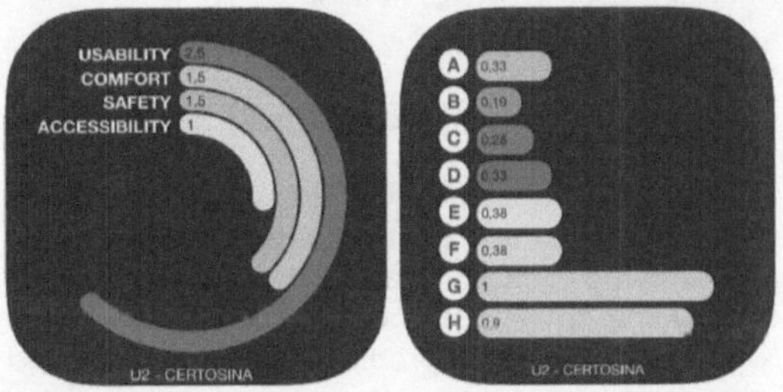

Figure 7. Cross-reference evaluation chart completed for the case study.

This is due to the hypothesis of demolish an existing intermediate slab, built in late XIX century, in order to reconstruct the original internal spatiality of the church. The specific REV chart (Fig. 7) shows a specific prevalence of positive impacts of the planned transformations, as the result is a relevant upgrade of accessibility and future transformability (Fig. 8).

It is now under development the same process with a different set of simulated functional choices, in order to evaluate the sensitiveness of the indicators in case of comparative evaluation of different reuse strategies, and to complete the necessary fine-tuning of the defined parameters.

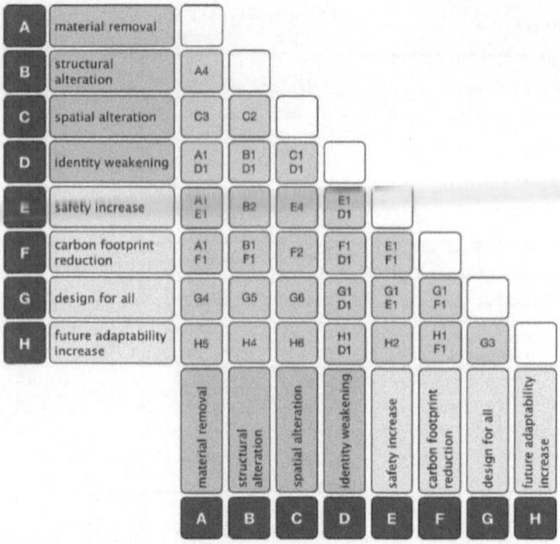

Figure 8. PAV and REV indicators specific to the case study. PAV shows a good functional adequacy level (Usability), average comfort and safety level and a low level of accessibility. REV shows a relevant impact of supposed design actions focused on improvement of accessibility (control variable G: design for all).

4 FURTHER DEVELOPMENTS

Further development will integrate economic parameters in the evaluation strategy assessment, and carrying capacity analysis for the functions to be settled in the building after intervention.

The study of an appropriate evaluation tool to compare different scenarios in term of financial relevance and impact will be stressed end developed, eventually applying value-analysis methodologies.

Main cost categories to be considered will be the restoration costs, the managing costs, and finally the maintenance related costs. Among these categories, at least the first one, more strictly connected with the restoration project will be analysed and defined.

A second level of investigation will also include the determination of specific thresholds of use, determined in relation to previously-measured functional hypotheses and the vulnerability and fragility of the building, coherently with the analysis methods of carrying capacity. The aim of this action is the determination of the maximum acceptable working load for the building by means of optimization procedures set between the uppermost operational limits of the asset (beyond which the human load compromises its conservation) and the lowest limits of use (below which the use of the asset becomes excessively limited). In this phase, therefore, it will be necessary to mediate between potentially conflicting actions of preservation and valorisation of the asset itself. Such an analytical phase is not addressed in this paper under the operational profile on account of its high specificity, but it is nonetheless fully integrated into the proposed model since the push towards fruition is one of the forces acting on the perturbed system and on its specific characteristics (physical, spatial, material). The same performance-based approach would allow the identification of the maximum threshold of users allowed, per unit of time, as a critical threshold, beyond which the system evolves towards a non-desirable perspective.

REFERENCES

Berkes, F., Colding, J. & Folke, C. (2003). Navigating social–ecological systems: building resilience for complexity and change. Cambridge: Cambridge University Press.

Brand, F.S., & Jax, K. (2007). Focusing the meaning(s) of resilience: resilience as a descriptive concept and a boundary object. Ecology and Society 12(1):23 [online] URL: http://www.ecolog yandsociety.org/vol12/iss1/art23/.

Buchanan, D. et al. (2005). No going back: a review of the literature on sustaining organizational change. International Journal of Management Reviews 7:189–205.

Carpenter, S.R. & Brock, W.A. (2008). Adaptive capacity and traps. Ecology and Society 13(2):40. [online] URL: http://www.ecologyandsociety.org/vol13/ iss2/art40/.

Carpenter, S.R. et al. (2001). From metaphor to measurement: resilience of what to what? Ecosystems 4:765–781.

Colucci, A. (2012). Le città resilienti: approcci e strategie, Pavia: Jean Monnet Centre Ed.

Folke, C., Colding, J. & Berkes, F. (2003). Navigating social-ecological systems: Building resilience for complexity and change, Cambridge: Cambridge University Press.

Folke C. et al. (2010). Resilience thinking: integrating resilience, adaptability and transformability, Ecology and Society vol.15, iss. 4.

Holling, C.S. (1996). Engineering resilience versus ecological resilience. Pages 31–44 in Schulze, P. editor. Engineering within ecological constraints. Washington, D.C.: National Academy Press.

Lavagna, M. (2008). Life Cycle Assessment in edilizia. Progettare e costruire in una prospettiva di sostenibilità ambientale, Milano: Hoepli.

Fabbricatti, K. (2013). Le sfide della città interculturale, Milano: Franco Angeli.

Jigyasu, R. et al. (2013). Heritage and resilience. Issues and opportunities for reducing disaster risks, Geneva: Global Platform for Disaster Risk Reduction.

Morandotti M. (2012). Contenuto vs. contenitore? Criteri predittivi di impatto e soglie di resilienza nella prospettiva del recupero sostenibile, IN_BO. Ricerche e progetti per il territorio, la città e l'architettura, 3(5), pp. 289–308.

Morandotti M. (2017). Resilience and memory. Notes for an approach to the sustainable transformability of immovable Cultural Heritage, in Besana, D. [RICH*] —Reuse and Improvement of Cultural Heritage, pp. 55–76. Roma: Aracne.

Innovative Built Heritage Models – Van Balen & Vandesande (Eds)
© 2018 Taylor & Francis Group, London, ISBN 978-1-138-49861-7

Review of standards for University built heritage spatial assessment

S. Lucenti & M. Morandotti
DICAr, Faculty of Engineering, University of Pavia, Pavia, Italy

ABSTRACT: In the early 1970s a new era of rapid expansion of universities in Western countries began. The rapid transition from elite university to mass university with an exponential increase of the number of students made the available spaces inadequate. Existing universities expanded in new peripheral areas and new universities were established. Many researches on the design of new settlements started. Today the situation has changed. We are experiencing a period of stabilization in university enrolment and many countries have undertook a process of spending review and of reorganization of public institutions. This involves the reduction of inefficiency in the use of spaces which can be a big challenge especially for European historic universities with heterogeneous building stocks. Thus a revision of the indices in literature and in handbooks and of regulatory standards for building design for university buildings aiming at evaluating the correspondence between functions and spaces in historic buildings and at minimizing the impact of the reuse on the built heritage.

1 INTRODUCTION

In assessing the compatibility between building and function—which is at the very basis of sustainable use of the built heritage—we need to value the available spaces and areas. Space management and occupancy optimization involve various aspect such as the typology (e.g. adaptability of the building's circulation plans, …), the time schedule (e.g. combined or temporary use, …) and the space standards (to plan and to manage the use of the buildings). The paper focuses on space standards, which are considered as indices for the assessment of the proper occupancy rate—aiming at avoiding a deep transformation of the building—rather than a rigid prescriptive tool.

In order to obtain a correct assessment of the heterogeneous building stock—such as the built heritage of universities (particularly historic ones)—we need to define standards which could be used both on historic and on modern buildings.

A challenge is to discuss and upgrade the standards developed in 1970s—for new University settlements—in order to use them effectively also for the assessment of historic buildings.

2 THE FIRST PERIOD OF UNIVERSITY CONSTRUCTION BOOM

At first Universities were funded in Europe in the Middle Ages. After alternating stages of expansion and consolidation they have become increasingly important institutions in the socio-cultural and economic reality of European cities. Different types of settlements can be distinguished, the oldest are the ones of universities distributed into the urban fabric (typical of southern European countries), urban campuses of universities in continental Europe, British colleges, North American campuses and the so-called European «university cities» (Lucenti, 2017).

A turning point in the development of universities in the western countries took place between the 1960s and the 1970s—after Students Protests—University passed from an elite University to a mass University. For instance at the University of Pavia (Italy) in less than 10 years the number of enrolled students increased dramatically. It passed from 3.000–4.000 students in 1965 to 15.000 students in 1974, see Fig. 1. In 10 years it had increased of almost the 400% (Lucenti & Zamperini, 2015).

Suddenly the available spaces in the building stocks of the existing Universities were totally

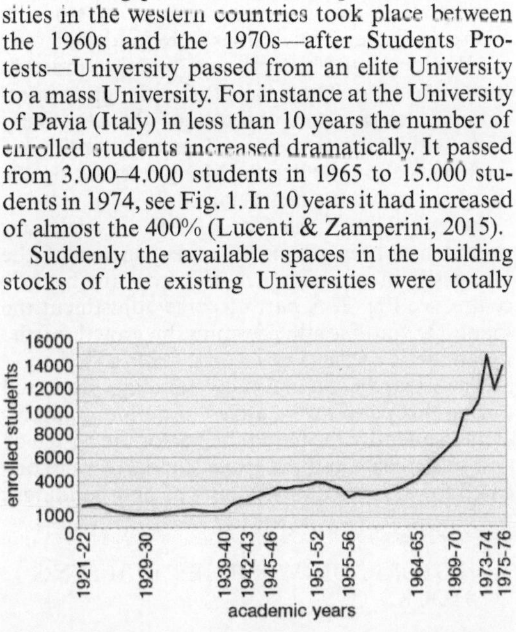

Figure 1. Number of students enrolled at the University of Pavia between 1921 and 1976, Lucenti & Zamperini (2015)—modified.

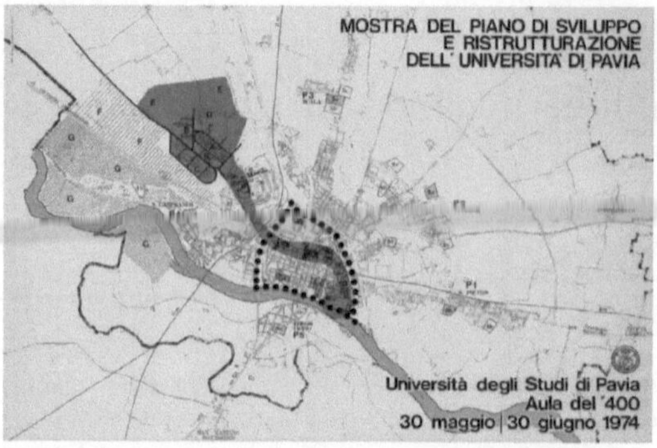

Figure 2. New external areas according Piano De Carlo for the expansion of the University of Pavia, Lucenti Zamperini (2015)—modified.

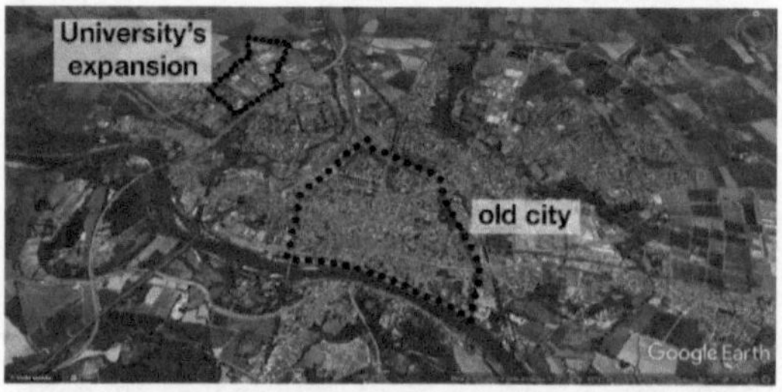

Figure 3. Aerial view of Pavia: the historic centre, University expansion according to Piano De Carlo, image from Google Earth—modified.

inadequate to the increasing number of students and so many new Universities were funded all over the World and the historic ones were expanded in peripheral areas. These facts fostered a great international debate about the role of the University itself within the Society and great attention of Architects about its design. Many research on the design of new settlements started.

In the 1970s in the City of Pavia a visionary project had been promoted of expansion of the University in an area north-east from the historic centre, see Fig. 2. A part of some adjustment the Piano De Carlo is still affecting the growth of the urban fabric of Pavia nowadays, see Fig. 3.

The researches on university buildings promoted during this period were aimed at defining parameters to quantify the requirements for the design of new settlements and the space standards were oriented towards the dimensioning of new buildings.

3 HISTORIC UNIVERSITIES' BUILDING STOCKS

Numerous socio-cultural and economic factors have influenced the development of historic universities' built heritage in various stages. In the end it became a very heterogeneous building stock, that incudes buildings:

- from different ages (which implies a different perception of their cultural value and different restriction in the possibilities of intervention: restoration, refurbishment, ...);
- with different typology (they were often reused as Universities after having been built for other functions and so the internal layout has been modified during centuries);
- with different constructive characteristics (e.g. the thickness of structural walls and the width of windows...).

It is also necessary to specify a distinction between historical universities hosted in urban buildings where is a strong integration between university settlement and the city and other more spatially-defined settlements which—although the growth of urban centres has often incorporated them into the urban fabric—still maintain a more "autonomous" functioning.

Within the building stock of Universities are settled spaces and buildings both for teaching and research and for their administration (see Fig. 4). All of them are changing rapidly and they need to be often upgraded to meet the constantly-changing needs expressed by users.

Different reflections have to be made for university facilities that—especially in the case of settlements widespread in the urban fabric—are often shared with the city (often this kind of university relies on a system of infrastructure and services not of its own exclusive relevance, such as: transport infrastructures, sports and commercial equipment, ...) (Cuscito & Cervini, 2006).

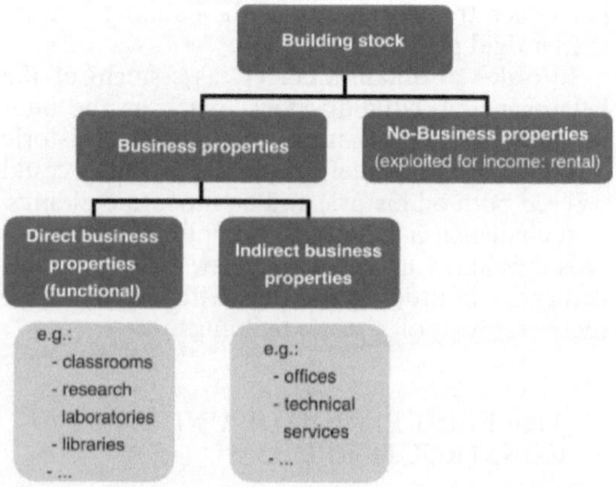

Figure 4. Classification scheme for universities' building stocks, further subdivided into space used for educational purposes (bottom left) and educational facilities (bottom right), Del Gatto 2010—modified.

4 THE NEW NEEDS AND THE DEMAND FOR REORGANIZATION

Nowadays we discuss about the changes in the way of learning and about the change of learning spaces. Some authors propose to open University to city and to create new places for learning—within the city itself (López-Toribio Moreno, 2015)—others talk about the virtualization of the educational models (Kiib, 2006).

Meanwhile we need more and more flexibility in workplaces too and the teleworking is spreading (Ciaramella & Bellintani, 2015).

In the last decades a transition period has started also for working tools (e.g. from typewriters to personal computers, from paper archiving to digital archiving, …) and consequently the space for a specific work activity has been reduced while—to guarantee security and accessibility in public buildings—the rules about the dimensioning of circulation spaces and escape route have become more and more strict.

In addition the awareness of the importance of University built heritage all over the world has achieved a deep meaning for «historic» Universities: it has a symbolic value (closely linked to the identity of university communities and cities) and also a marketing value in attracting students and researchers (somehow having historic building stock has become part of the brand of a University) (Bulotaite, 2003; Merchant et al., 2014). Furthermore after the deindustrialization process, cities in Western countries still maintain the role of centres for the production of culture (Dilorenzo & Stefani, 2015). So these historic building complexes have become irreplaceable both for their symbolic value and for their strategic location inside the urban fabric and the functions inside them can't be easily-transferred even if they don't satisfies users' requirements anymore.

After 2008 financial crisis spread worldwide there has been a period of economic stagnation and a reduction of the Public Spending. Between the 2008 and today in Italy the spending review has led to a reduction of about the 20% on the investments on Universities (ANVUR, 2016).

The Government has also required an optimization of the use of the building stocks of Public Administration, aiming at fostering a more efficient use of available spaces (which has become to be considered an important asset not to be wasted).

This aim for a better use of available floor areas is a quite obvious requirement for privates but is rather new for the Public Sector and has become a fundamental after the reduction of investment.

The adoption of their own building stock management tools distinguishes Anglo-Saxon-influenced universities from those of southern European countries. It depends on multiple concomitant factors such as: greater administrative autonomy of private universities, greater cultural attention to the topic of maintenance, the more circumscribed settlement type in delimited areas (i.e. campus and college) that allows a more effective control of buildings.

In fact in the last years many British and north American universities have adopted their own guidelines both for facility management and real estate (e.g. the guidelines of the Penn University, Philadelphia (USA), the Space Management Policy 2010 of the University of Strathclyde, Glasgow (UK), the Space Planning Guidelines by the Australasian Association of Higher Education Facilities Officers).

4.1 *Present situation*

In the last ten years we have been experiencing a period of stabilization in university enrollment. Thus there are few new university settlements and an increasing number of interventions upon existing buildings. However reuse and reorganization of existing buildings have to deal on the one hand with the emergence of new functions and the continuous request for an upgrade due to the rapid change of users requirements, and on the other with the increasing attention to environmental and economic sustainability of interventions and with the awareness of the cultural and economic value of built heritage, (Morandotti, 2016).

The diagram in the Figure 5 shows the number of students enrolled at the University of Pavia between 2006 and 2017. Over the past 10 years the number of enrolled students has stabilized at around 22,000, with a slight increase during the last academic year (see Fig. 5).

Furthermore after the 2008 financial crisis in many countries the public sector has undergone a deep spending review. This has led to a reduction

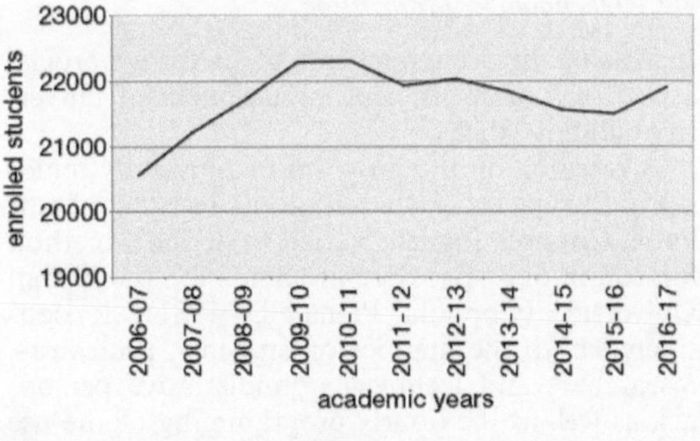

Figure 5. Students enrolled in the University of Pavia between 2006 and 2017, scheme by S. Lucenti, datas from MIUR (http://anagrafe.miur.it).

	P	U		a_x	q_x	k_x	Stud.	Note
	min	min	MAX	[m²/ab]	[m²/ab]	[m³/ab]	Prof.	
M Arch.	100000			20-25				A_x min 3-10 ha
N.M.Arch.		5000	30000					
faculty of science					50	150*		*for students, professors and staff
faculty of humanities					75-100	250*		
faculty of medicine					150-200	500*		
MPH		20000	25000			40-100	7-11	
Rossetti	100000	100000			20-25			A_x min 3-10 ha
French norms					26			I_u = 1.7 %
English norms		3000					8	
ORL-ETH				1.5-3				
German norms		6000	10000		4-4.5	humanities		50 ha per
					15-18	science		
European norms				0.43	20			
American norms		25-30000						
Unesco								
Architecture					45*			*for students, professors and staff
Medicine, Engineering, Natural science, Music					40-35*			
Faculty of humanities					200*			

Figure 6. Comparison between various urban planning standards for University settlements, table from Mercandino (2009)—modified.

in public investment that has hit in particular the non-core business branches (such as facility management and real estate at public universities (De Medici & Senia, 2015)) from which the need to reduce the under-use of buildings has arisen (Lucenti, 2017).

A possible strategy is to improve the efficiency of the exploitation of available spaces. The reorganization of interior spaces—what (Isakova, Morgun & Evtushenko-Mulukaeva, 2013) mention as *the adaptive method* and as *the method of internal reconstruction* – represents for universities hosted in historic building stocks an opportunity for an upgrade (i.e. to meet users' new needs, to improve security, …) and for an assessment of sustainability of use—that is the actual compatibility between functions and spaces—which guarantees the maximum preservation of the built heritage.

Hence—for a proper occupancy load assessment—a review of space standards and a discussion on their applicability to historic buildings are needed.

4.2 *Occupancy optimization*

Increasing the occupancy rate of spaces is a crucial aspect of a more efficient management of university building stocks.

A research on the program of university building in Europe conducted in the 1970s by Professor Paola Coppola Pignatelli tried to define a method for computing space requirements for the Italian University (Coppola Pignatelli, 1981). It considered both the number of students, professors, researchers and technical administrative personnel, as well as the hourly programming of the use of buildings.

The time factor is a very important element that needs to be taken into account through a careful scheduling of classroom occupation—even through the implementation of an appropriate algorithm (Abdelhalim & El Khayat, 2016) – and by assessing spaces occupancy rates of students, professors and technical and administrative staff (Den Heijer, 2011).

In recent years thanks to the establishment of the role of facility management—even in the public sector—some of the universities have begun to optimize the use of spaces. In this sense, on the one hand, settlements widespread in the urban fabric allow for a more direct (in some cases symbiotic) relationship between historic universities and cities (thus reducing the typical emptying effect of campuses in the closing time), on the other hand, universities hosted in historic buildings are facing the challenge of implementing management models for a built heritage with strong heterogeneity and complexity (Fig. 6).

5 REVIEW OF AVAILABLE SPACE STANDARDS FOR UNIVERSITY BUILDINGS

It is possible to distinguish four typologies of space standards for University design:

1) standard available in scientific literature;
2) standard in state building regulation;
3) standard from Real Estate market;
4) Universities' guidelines space standards.

5.1 *Space standards in scientific literature*

The standard available in scientific literature can be subdivided into two different levels (see Fig. 7):

1. Urban planning level (focused on the evaluation at large scale of new settlements);
2. Standards focused on the building scale (that is on the space planning requirements).

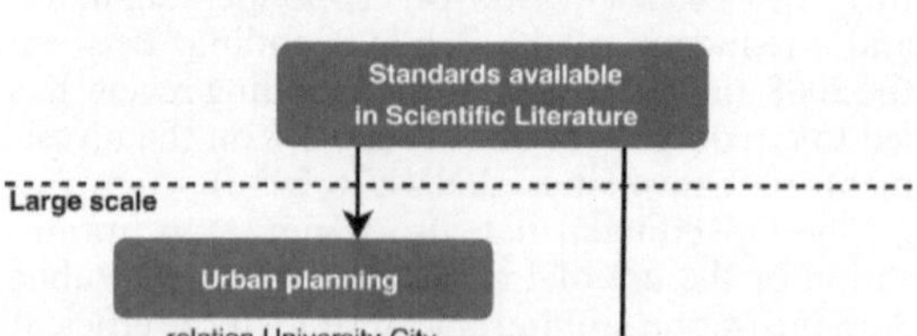

Figure 7. Different scales of standards available in scientific literature, scheme by S. Lucenti.

They both propose Gross Areas defined as the sum of: internal areas, walls and structures, installations, spaces for circulation and other spaces.

The space standards in many handbooks and manuals of architecture refer to the results of the research from the 1970s and 1980s, focusing mainly on new buildings rather than on the topics of reuse and reorganization of existing buildings.

The already cited research led in 1970s by professor Paola Coppola Pignatelli upon the planning of University buildings in Europe tried to make a state of the question of that period and to define a method to evaluate the spatial need for Italian University. Among Italian case studies they analysed the Piano De Carlo for the University of Pavia (1974) as a good practice that was being put into practice at that time (see space standard in Fig. 8).

5.2 *Space standards in state building regulation*

Although in many countries university and research are matters of direct relevance to the state, we can affirm that in general there is a lack of specific governmental or local regulation about university buildings, perhaps because of the great administrative autonomy (and self-organization) that has been granted historically to university institutions.

For instance in Italy and in Spain there are no specific state regulations on university buildings but reference is made to requirements for similar functions which may also have significant deviations (e.g. for large classrooms they refer to buildings for public show, for chemical and mechanical laboratories they refer to the rules for industrial buildings, for departmental spaces they refer to office buildings, …).

Since the early 2000s in Italy the government have tried to foster the reorganization aimed at greater efficiency in the use of space in public offices buildings with a set of laws. A first attempt was made with the decree *D.M. 14 marzo 2001* and subsequently with the decree *D.L. 6 luglio 2012, n. 95,* after whom the *Agenzia del Demanio* (i.e. the State Property Office) provided the Public Administrations with instructions on the circular *16 luglio 2012, Prot. n. 2012/20494* (see Fig. 9) (Lucenti, 2017).

The standards published in the *D.M. 14 marzo 2001* consider the Gross Areas, which includes: the

Space standard	Engin.	Math.	Bio.	Pharm.	Econ.	Polit.	Literat. Philos.	Law
	[square meters]							
students	19.5	7.0	16.0	50.0	6.6	14.3	13.1	12.3
professors, researchers	145.0	192.0	102.0	102.6	325.4	66.7	61.0	76.6

Figure 8. Resuslts of the space standards analysis of the University of Pavia in the 1970s by Professor Paola Coppola Pignatelli, table from Coppola Pignatelli (1980)—modified.

Standards for public offices		
job position	Maximum number of peopleper room	m² per person
Manager	1	min 25,3
		MAX 28,3
Functionary (levels from 7th to 9th)	3	min 13,3
		MAX 21,3
Employee (levels from 4th to 6th)	3	min 9,0
		MAX 12,0

[D.M. 14 marzo 2001]

Change in state regulation

Space standard	
case of application	m² per person
new buildings and complete renovation	min 12,0
	MAX 20,0
reorganization of existing buildings	min 20,0
	MAX 25,0

[D.L. 6 luglio 2012, n. 95]

Figure 9. Comparison of various Italian regulatory indication in recent years, scheme by S. Lucenti.

actual spaces for work (i.e. offices), all the complementary spaces (i.e. meeting rooms, libraries, archives, cafeterias/canteen, …) and also spaces for circulation and installations (i.e. corridors, stairs, elevators, toilets, boiler room, …). It is also important to notice that those standards still mark a subdivision by a hierarchical order starting from lower level employees up to top managers.

In 2012, the *Agenzia del Demanio* provided two different ranges of values by eliminating the previous hierarchical division in favour of a distinction in two categories of intervention:

1. new constructions and integral renovations (12–20 m^2/pers);
2. reorganization of existing buildings (20–25 m^2/pers).

5.3 *Space standards from Real Estate market*

On the other side the assessments operated by the Real Estate market are based on economic drivers and take into account the net rentable areas and the indeces of efficiency.

Corporations are fostering new ways of using available spaces inside a building based on a more flexible way of working and interconnected interior spaces that are more than offices. That's why in the last 15 years areas a reduction of the net workplace area has been observed (see Fig. 10) and actually this trend has been assumed also by the 2015 *International Building Code* which considers a minimum area of 100 square feet/employee (9.29 m^2) for the working space.

Particularly Real Estate's building stocks assessments are based on the evaluation of some indeces of efficiency:

- Primary circulation/NRA (in case of building typology with corridors it's good if = 0.15–0.20)
- Owner efficiency = NRA/GIA (good if > 0.80; bad if < 0.70);
- Tenant efficiency = NUA/NRA (good if > 0.85; acceptable if = 0.80–0.85; bad if < 0.75).

With NRA = Net Rentable Area, GIA = Gross Internal Area, NUA = Net Usable Area (Ciaramella & Bellintani, 2015).

5.4 *Universities' guidelines space standards*

A further category of space standard is the one defined by universities themselves. As already mentioned in paragraph 4, many universities are providing their own internal guidelines on the matters of facility management and real estate. These include space management rules and space standards.

From an international review of current guidelines of universities emerge that many institution consider: for department buildings a workspace Standard Area of about 12.00 m^2/pers, while for learning spaces the standard area is about 2.00–2.50 m^2/stud.

The Space Planning Guidelines, Edition 2 published by the *AAPPA. Australian Association of Higher Education Facilities Officers* reports: «For a broad 'rule of thumb' planning parameter, 15 m^2 GFA/EFTSU would be considered an Australasian average for university space. This ratio covers all space on a campus except for student housing» (Stephenson, 2006). With GFA = Gross Floor Area, EFTSU = Equivalent Full Time Student Unit, which is «a value representing the student load for a unit of study or part of a unit of study...» (Stephenson, 2006), so the dimensioning of campus spaces are referred directly to the number of students and to their study load, while considering the utilization rate.

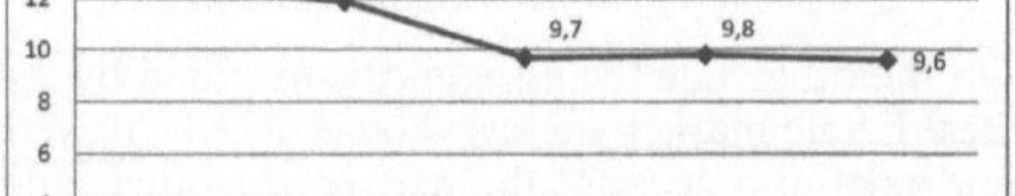

Figure 10. Reduction of net area [m^2/desk] in the UK during the last 10 years, diagram from (Ciaramella & Bellintani, 2015).

6 REVIEW PROCESS

The aim is to define standards applicable in case of evaluation of the spaces of the building complexes of historic universities both in the case of post occupation evaluation or in case of change management.

Starting from the hypothesis of a certain functional program, the need for space was defined, with a formula like:

DESIRED # OF OCCUPANTS/LOAD FACTOR = SIZE OF REQUIRED HOST STRUCTURE, Wong (2017)

With LOAD FACTOR = 1/Standard Area = number of occupants/area [pers/m^2].

Then evaluating the deviation of different results by varying the standards (in particular between those of the scientific literature and those of the state building regulations).

Later an analysis has been done of plans and of the layout scheme for the evaluation of the indices for buildings' efficiency assessment from the real estate market.

A further step has been added, i.e. the comparison between historical (pre-industrial) and modern building (characterized by dimensional standardization and industrialization of the constructive process) for calculating the deviation [%] due to typological factors (due to the reuse of buildings with different initial use) and constructive (due to the different incidence of structures and internal walls on the gross area).

7 THE CASE STUDY

The current research origins from the demand of reorganization of the University of Pavia. Initially the case study of *Palazzo San Tommaso* – the new house of the Department of humanities—have been analysed (see Fig. 11).

7.1 *Palazzo San Tommaso*

Palazzo San Tommaso is a large building complex in the old town of Pavia. Now it belongs to the University of Pavia but previously it had been a convent, then a seminary and then a barracks.

In last years, within the renovation of the whole complex the University has begun also the

Figure 11. Aereal view of the Central Palace of the University of Pavia (on the left) and Palazzo San Tommaso, Department of humanities (on the right), image from Google Earth—modified.

reorganization of the spaces of the Department of humanities (see Fig. 12), which include: some classrooms, a new large library and the offices for the professors and the staff. We are carrying out a two-level verification of the compatibility between function and spaces aiming at reducing the impact of the reuse process on the built heritage. We are focusing more specifically on the reorganization of spaces assimilable to offices.

The first verification between space requirements (in terms of Gross Area defined from the number of users) and the spaces available for the offices (i.e. not taking into account the spaces allocated to the classrooms and the new humanities library) has demonstrated the possibility to host the whole Department into the available spaces of the historical complex. To estimate the range of acceptable values of the space demand for the minimum values the standards of handbooks have been considered while for the maximum values Italian laws have been considered.

The analysis results of the first verification has shown that globally there is a surplus of available space between 40÷90% (see Fig. 13), thus the possibility of increasing the initial occupancy rate. While the verification for each discipline cluster has highlighted which one could be improved (see Fig. 14).

Then a second verification on the compatibility between the organization structure (which lead to a space model, considering the specific needs of the users) and the typology of historical buildings (which is different from modern ones) has followed. In fact in the evaluation of the spatial/functional compatibility it has to be considered both the building typology (with: large corridors, serial rooms, great internal height, great halls, …) and the characteristics of the construction (i.e. great thickness of the structural and internal walls, …).

At the present time we are analysing the first results of the second verification. They seem to suggest that the rigid department organizational

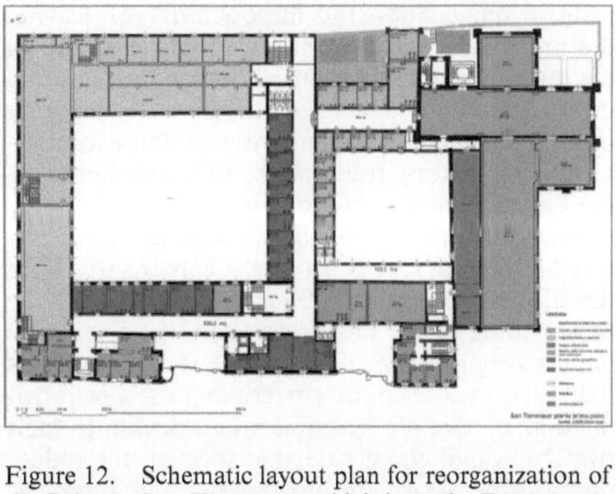

Figure 12. Schematic layout plan for reorganization of the Palazzo San Tommaso, which host the Department of humanities of the University of Pavia, scheme by S. Lucenti.

Occupancy verification of the whole department					
Department		space demand [m²]	available space [m²]	verification [✓/ ✗]	deviation [%]
Humanities	min.	3315,00	5269,94	✓	37
	MAX.	4878,00			93

Figure 13. Tables of the first verification of the space demand for the whole department of humanities.

Occupancy verification by disciplinar cluster					
Discipline cluster		space demand [m²]	available space [m²]	verification [✓/ ✗]	deviation [%]
Antiquities	min.	738,00	1209,17	✓	39
	MAX.	1160,30			4
Literature	min.	744,00	906,30	✓	18
	MAX.	1046,20			-15
Geography	min.	774,00	1385,44	✓	44
	MAX.	1216,90			12
Linguistic	min.	507,00	653,93	✓	22
	MAX.	706,90			-8
Foreign languages	min.	447,00	857,38	✓	48
	MAX.	621,70			27
Secretary	min.	105,00	212,72	✓	51
	MAX.	126,00			41

Figure 14. Table of the first verification of the space demand for each discipline cluster.

structure and the rigid typology of the building tend to reduce significantly the deviation between space demand and available space.

8 CONCLUSIONS AND PROPOSAL OF DEVELOPMENTS

In the light of the growing importance of the protection and enhancement of the built heritage of historic universities, the impossibility of leaving the prestigious institutional headquarters and, at the same time, their continuous need for upgrading both to meet the new needs of the ever-changing users' requirements and to respect the evolution of the regulatory framework, the assessment of the compatibility between function and spaces is a fundamental aspect of the sustainable use of historical building stocks. In fact it can ensures both the efficiency of the public institutions and the preservation of the buildings. From these requests has its origin a need for an update of standards that take into account current trends: the introduction of flexible learning and working models (which require equally flexible spaces), the reduction of office space, the growing need for Intervening on historic buildings (to reduce the cost of use of buildings by improving the efficiency of exploitation of available spaces, also reducing the waste of non-urbanized land) and the need to assess the sustainability of such interventions on existing buildings by ensuring the maximum compatibility between function and space (ensuring at the same time maximum conservation of the built heritage).

Further results of this research will be: the verification of the adequacy of benchmarks used in the assessment of the occupancy of historic buildings, the verification of the effectiveness of indeces for occupancy efficiency assessment of historic buildings and the definition of correction coefficients to update the space standards (which often derive from the research from the 1970s).

REFERENCES

Abdelhalim, E.A. & El Khayat, G.A. (2016). Utilization-based Genetic Algorithm for Solving the University Timetabling Problem (UGA). *Alexandria Engineering Journal* 55(2): 1395–1409.

Cecchi, D. (ed.) (2016). Rapporto biennale sullo stato del sistema universitario e della ricerca. Roma: ANVUR – Agenzia Nazionale di Valutazione del sistema Universitario.

Bulotaite, N. (2003). University heritage—an institutional tool for branding and marketing. *Higher Education in Europe*, 28(4): 449–454.

Coppola Pignatelli, P. (ed) (1981). Programmazione dell'edilizia universitaria in Europa Ricerca di un metodo per la definizione del fabbisogno spaziale per l'università italiana. Roma: GREIS, Italposte.

Ciaramella, A. & Bellintani, S. (2015). Spazio ufficio, Programmare, progettare e gestire l'ufficio contemporaneo. Santarcangelo di Romagna: Maggioli Editore.

Cuscito, L. & Cervini, R. (2006). The management of student related facilities in higher education. In Pietroforte, R., De Angelis, E. & Polverino, F. (eds.) *Construction in the XXI century: Local and global challenges, Proc. Int. Symp. Joint CIB W065/W055/W086, Rome, 2006:* 1–11. Naples, Edizioni Scientifiche Italiane S.p.a.

De Medici, S. & Senia, C. (2015). Ottimizzazione dei servizi di manutenzione per l'edilizia universitaria. *Techne* 9(5): 267–273.

Del Gatto, M.L. (2010). Gestire gli edifici pubblici teoria e pratiche. Santarcangelo di Romagna: Maggioli Editore.

Den Heijer, A.C. (2011). Managing the university campus Information to support real estate decision. Delft: Eburon Academic Publisher.

Dilorenzo, P. & Stefani, E. (2015). Università e città, Il ruolo dell'università nello sviluppo dell'economia culturale delle città. Roma: Fondazione CRUI.

Isakova, S.A., Morgun, N.A. & Evtushenko-Mulukaeva, N.M. (2013). Space-Planning Modernization Methods for Educational Buildings of the Southern Federal University. *Middle-East Journal of Scientific Research* 17(10): 1412–1418.

Kiib, H. (2008). Innovative Universities and the Experience City. In Fink, F. K. (ed.), *Proc. SEFI 2008, 36th Annual Conference, European Society for Engineering Education, Aalborg, 02–05 July 2008:* 1–10. Aalborg: Sense Publishers.

López-Toribio Moreno, R. (2015). Caso especifico de la E.T.S.A.G. en el Barrio de El Realejo. Granada: Escuela Técnica Superior de Arquitectura de Granada.

Lucenti, S. (2017). La gestione sostenibile del patrimonio edilizio delle Università storiche. Sustainable management of the built heritage of historic Universities. In *Proc. symp. Colloqui.AT.e. 2017 Demolition and Reconstruction* ? Ancona, 28–29 September 2017. In press.

Lucenti, S. & Zamperini, E. (2015). Urban symbiotic architecture in Pavia The origins of Giancarlo De Carlo's Building Development Plan for Pavia University. In C. Gambardella (ed.), *Heritage and Technology. Mind, Knowledge, Experience; Proc. XIII International Forum of Studies 'Le Vie dei Mercanti', Aversa and Capri, 11–13 June 2015:* 1446–1455. Napoli: La scuola di Pitagora editrice.

Mercandino, A. (2006). Urbanistica tecnica Pianificazione generale: 300–302. Milano: Il Sole 24 ORE.

Merchant, A., Rose, G.M., Moody, G. & Mathews, L. (2014). Effect of university heritage and reputationon attitudes of prospective students. *International Journal of Nonprofit and Voluntary Sector Marketing* 20(1): 25–37.

Morandotti, M. (2016). Università e città: sfide e opportunità per uno sviluppo sostenibile. *Città in Controluce* 28(20): 1–9.

Stephenson, D. (ed.) (2006). Space Planning Guidelines, Edition 2. AAPPA. Australian Association of Higher Education Facilities Officers.

Wong, L. (2017). Adaptive REUSE Extending the lies of Buildings. Basel: Birkhäuser.

Innovative Built Heritage Models – Van Balen & Vandesande (Eds)
© 2018 Taylor & Francis Group, London, ISBN 978-1-138-49861-7

The historic urban landscape practice in Latin America, Cuenca as a case study

M. Siguencia Avila
Universidad de Cuenca, Cuenca, Ecuador
KU Leuven University, Leuven, Belgium

S. Vintimilla Andrade
Universidad de Cuenca, Cuenca, Ecuador

ABSTRACT: Nowadays, based on the 2011 UNESCO Recommendation on Historic Urban Landscapes (HUL), the need to implement efficient tools to ensure urban conservation, is crucial in the heritage agenda. In Latin America the change of millennium brought about a dramatic urban expansion that exceeds the capacity of historic centres, especially those included on the World Heritage List (WHL) bear values that can be threatened. The HUL notion which is raised in light of new architecture inserted in historic contexts, tackles a broad understanding of urban heritage and encourages the use of four groups of tools and a six-step action plan. Cuenca in Ecuador provides a framework on the use of such tools through meetings, projects and publications, while initiatives in Latin America manifest implementation processes. This study reports the results of a comparative analysis between the methodology used in Cuenca and the rest of cities in Latin America. This provides insights of a first evaluation of case studies in this region and their input for further application.

1 INTRODUCTION

The 2011 Recommendation on Historic Urban Landscapes, hereby referred to as the *Recommendation*, is the latest notion adopted by UNESCO proven to be an innovative method for management and heritage preservation of historic cities (UNESCO 2011). Within this approach, backed during the 2005 Vienna Memorandum, both cultural and natural values are integrated. These values shape multiple so-called layers of information in urban territories that need to be managed through tools integrating current and new ways of heritage management at local, national and transboundary level (Bandarin *et al.* 2011). The *Recommendation* explains (UNESCO 2011) (UNESCO 2011) HUL as "*the urban area understood as the result of a historic layering of cultural and natural values and attributes, extending beyond the notion of 'historic centre' or 'ensemble' to include the broader urban context and its geographical setting*"(UNESCO 2011).

Although early documents propose integrative methods to conserve heritage areas (Lemaire 1915, Geddes 1949, Whiteh and 2007), four groups of tools, namely knowledge, regulatory, financial and civic engagement tools, are proposed within the HUL notion (O'Donnell *et al.* 2012). Additionally, a six-step action plan is available for voluntary implementation which includes: mapping resources, reaching consensus, assessing vulnerability, integrating development frameworks, prioritising actions and establishing partnerships. These tools and methods intend to support processes facing the world's rapid urbanization for the coming thirty years (UN-Habitat 2013). This phenomenon brings an adaptation to new land-uses which are not always compatible with spaces bearing heritage values.

Cultural and natural values in urban settlements are universally recognised by UNESCO. Currently, the World Heritage List (WHL) holds 1052 cultural and natural properties, of which 138 are located in Latin American and the Caribbean region (UNESCO 2017). The nature of values in this area—where nature, landscape and intangible aspects are relevant—increasingly engages multidisciplinary processes for heritage management (Rojas 2002). During the last three decades, cities in Latin America have made efforts to better conserve historic areas through rehabilitation and planning actions (Steinberg 1996, Rojas 1999). However, these experiences are generally limited to historical bounds and conceived in technical grounds. Since 2012, Cuenca in Ecuador located in the Andean Region of South America is part of the WHL since 1999. Since 2014, the city has executed an academic-led initiative to adopt the tools proposed in the HUL notion supported by local, national and international institutions. Cuenca is the basis for a comparative analysis on case studies implementing HUL in the region, with special attention to World Heritage (WH) properties. Methods used in Cuenca are in contrast to those gradually delivered in Latin American cities.

2 WORLD HERITAGE AND HISTORIC URBAN LANDSCAPE

2.1 *World Heritage in Latin America*

Concern about heritage has shifted in the last decades from object-based categories to value-based approaches for heritage management (Vandesande 2017). This change has continuously evolved parallel to the concept of heritage, which started from protection of artworks and nowadays includes both tangible and intangible elements in local, national, transnational and worldwide scales. The intention to conciliate values related not only to single monuments but also to complexes of buildings, urban ensembles, cities and landscapes, is a long-standing tradition in history (Veldpaus and Pereira Roders 2013a).

Charters and recommendations affirm a context's relevance when analysing values embedded in monuments. The 1931 Charter of Athens, the 1964 Venice Charter, the 1976 UNESCO Recommendation concerning the Safeguarding and Contemporary Role of Historic Areas, and the Vienna Memorandum, are among the documents demanding the leading role of historic areas (UNESCO 2005).

UNESCO has played a decisive role for the recognition of worldwide heritage through the 1972 Convention concerning the Protection of the World Cultural and Natural Heritage. This document defines the typologies under which natural and cultural elements can be inscribed on the World Heritage List (WHL). This representative group of properties bear Outstanding Universal Value (OUV), demonstrating the local interest of States attached to the 1972 Convention to recognise heritage goods on their territory.

Within the 1052 properties inscribed on the WHL up to 2016 in 165 State Parties, 78 percent represents cultural properties while 19 and 3 percent correspond to natural and mixed properties respectively (UNESCO 2017). In the five regions recognized by UNESCO, Europe holds the highest number of inscriptions (498), followed by Asia and Latin America (243 and 138 properties). The regions with most properties in danger are Arab States and Africa with 21 and 17 properties out of the total 55 properties endangered (Figs. 1–2).

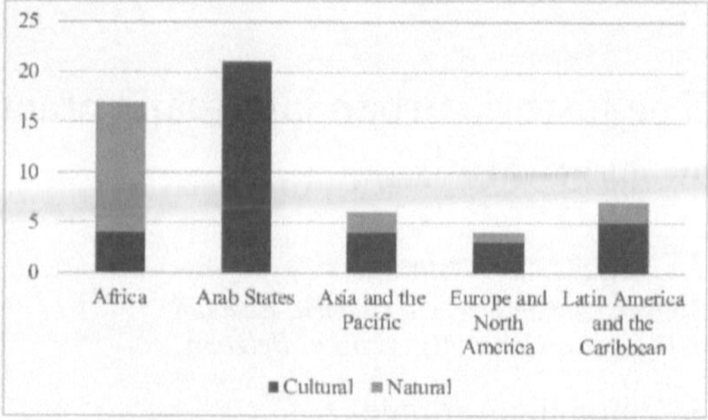

Figure 1. World Heritage by region.
Source: (UNESCO 2017).

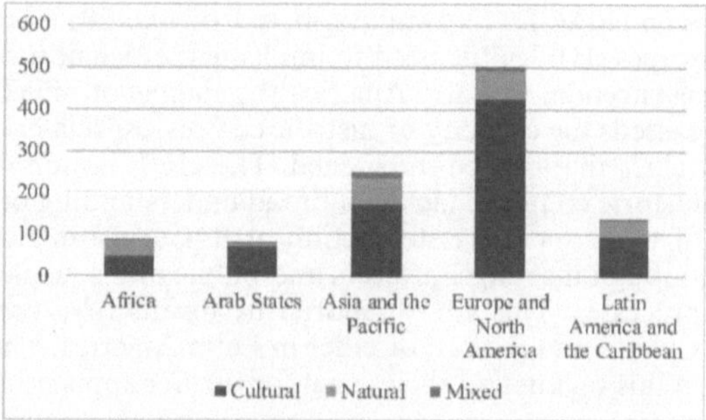

Figure 2. World Heritage in danger by region.
Source: (UNESCO 2017).

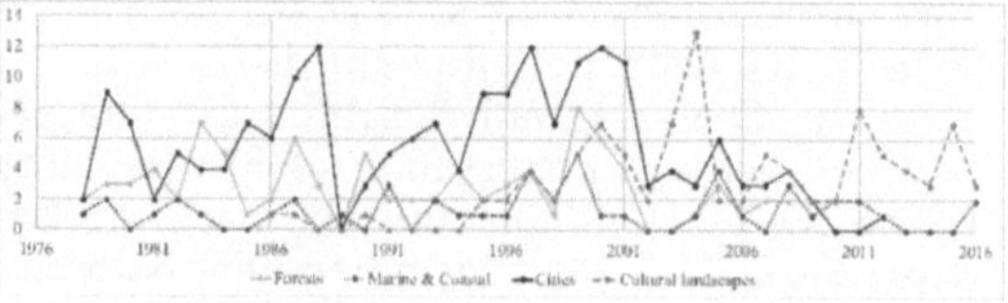

Figure 3. Timeline of inscriptions on the WHL in subcategories.
Source: (UNESCO 2017).

To better understand the diversity of the WHL, UNESCO presents four subcategories of World Heritage properties: marine and coastal, forests, cities and cultural landscapes. However, only 431 properties (40 percent of the total) have been categorised in these four groups. Figure 3 shows the predominant number of cities and cultural landscapes throughout time. From the 191 properties categorised as cities, Europe holds 110 (57 percent) followed by Latin America and Arab States with 37 and 21 cities respectively.

Latin America is formed by 28 State Parties with 138 properties, representing 13 percent of the total number of worldwide inscriptions on the

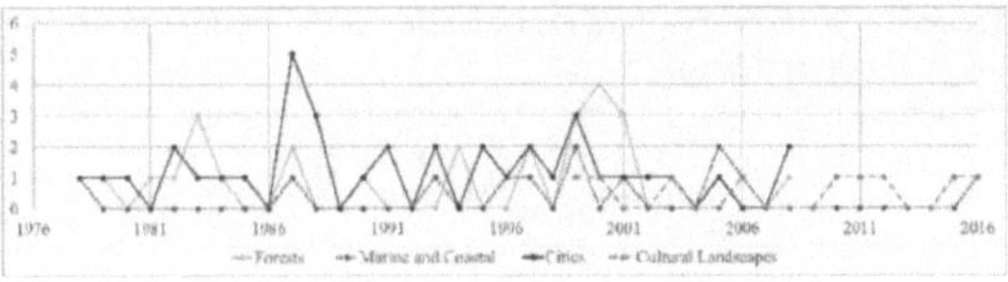

Figure 4. Timeline of Latin America inscriptions on the WHL in subcategories.
Source: (UNESCO 2017).

WHL. Cultural, natural and mixed properties are distributed with 70, 26 and four percent respectively. In the timespan, since the formation of the WHL, the four subcategories show peaks in the number of cities in 1987 and forests in 2000 while cultural landscapes only appear in 1998 (Fig. 4). A number of 84 properties are subcategorised in this region, with a highest percentage of cities (44 percent), followed by forests, marine and coastal properties, and cultural landscapes (30, 14 and 10 percent respectively).

These figures show the important number of WH properties considered as cities and how the evolution of the definition of heritage has influenced their recognition. Initiatives to further identify urban heritage includes the research project entitled Protected Urban Planet, which launched in 2009 to promote information exchange on the evolution of worldwide protected urban areas (Verbruggen *et al.* 2014). In this project, all WH nominations were reviewed to trace the territory related to not only cities but also to monuments or places associated to urban settlements.

2.2 *The 2011 HUL recommendation*

Heritage conservation in urban areas is implied in many charters and recommendations during the previous century (Veldpaus and Pereira Roders 2013b). From its first nominations, the WHL contains cities bearing Outstanding Universal Value, designation that recognises a broad conception of heritage that goes beyond buildings. The 2011 UNESCO Recommendation on Historic Urban Landscapes, originated from ideas implemented in charters and documents concerning the protection of urban areas. This soft-law is available for States to adopt the approach dealing with both urban heritage and new development in cities. The accelerated rate of urban growth during the 21st century in several developing areas around the world, has brought pressures for historic cities and its context. The results of such growth include environmental distress, massive tourism, and identity loss (O'Donnell *et al.* 2012).

HUL tackles these issues through the identification of a layering and interconnection of natural and cultural, tangible and intangible, international and local values in any city (Jokilehto 2010, UNESCO 2011). To this aim, a set of four specific tools is suggested which refers to knowledge, regulatory, financial and civic engagement. Despite not being adopted in the final document in 2011, the draft proposal of the *Recommendation* on HUL endorsed a six-step action plan for its implementation. These steps as well as the mentioned tools are further reviewed in the coming sections.

3 HUL IMPLEMENTATION IN LATIN AMERICA

3.1 *Urban heritage conservation in Latin America*

Rojas & de Moura Castro (1999) argue the existence of two phases of urban heritage conservation in Latin America. The first one corresponds to an esthetical-based period of an elite group against real estate investments that destroyed historic buildings and led to the origin of protective legislative laws and regulations. The second consist of a value-based phase where governments establish institutions to properly apply the legislation and to promote monitoring of actions (Rojas and de Moura Castro 1999). In 2002, the same author stresses that preservation of Latin America's vast urban heritage is beyond the capacity of governments, which have many competing demands for their scarce resources and therefore an evolution to a third phase was needed. This phase contemplates social actors' participation where private, civil society, organizations and local communities become part of urban heritage conservation processes (Rojas 2002).

The principles stated by these authors do not greatly differ from the notion proposed by HUL. Despite being conceived in Europe during the 2005 Vienna Memorandum, prior to the adoption of HUL in 2011, two of the eight expert meetings took place in Latin America. The cities of Olinda and Rio de Janeiro in Brazil, were the venues for the Conference on Historic Urban Landscapes for America in November 2007 and December 2009 respectively (Bandarin and Van Oers 2012). Amongst the relevant contributions were the agreements on change as part of urban development, the need to clarify authenticity and integrity for historic urban landscapes as well as the attention that borders and buffer zones have for protected areas (Conti 2016). Nevertheless, Iberoamerican authors also find conflicts in the terminology and understanding of HUL. Conti (2012), reveals the slanted translation of the term HUL in Spanish, since 'urban' is sometimes placed behind historic and its connotation becomes an adjective to 'historic landscape' (Conti 2012). Early after the adoption of HUL, Lalana Soto (2011) states that 'landscape' is trendy and embraces many aspects that the term itself makes it difficult to apply and easy to distort (Lalana Soto 2011, Lalana Soto and Santos Ganges 2011).

These assumptions explain why the application of HUL is not yet a priority on the heritage agenda in Latin America. Despite the fact that the establishment of strategic planning processes is stressed as top priority in historic centres (Montes Lira 2001, Steinberg 2005, González Biffis 2013), a review of the cases which are committed to adopting tools for HUL implementation shows scarce applications. After the HUL experts' meetings held in Brazil, 16 cases in nine countries have undertaken actions to implement HUL. In most of them, academic contributions aim to understand the HUL notion for their specific context, while others hold conferences and events to discuss and share experiences. Few of them have introduced strong initiatives to fully implement HUL and make use of the four group tools and the six-step action plan in academic, social and governmental spheres.

3.2 *Case studies implementing HUL in Latin America*

Based on online and personal communication sources, it has been possible to trace initiatives involving on the implementation of HUL. Six-tenths of the cases registered activities such as: individual research studies, academic projects, governmental and organisation enterprises, publications besides local and international meetings. To structure the analysis, the activities are grouped into three main categories: meetings, publications and projects. The first two are generally short events and involve a limited number of stakeholders, while projects tend to encompass a wide range of disciplines and stakeholders. For the aim of this study, projects are selected in the following heading to understand in a detailed manner how the HUL six-step action plan as well as the four groups of tools are considered in the different projects.

As indicated in section two, when looking at WH in subcategories, after Europe, Latin America holds the largest number of cities. Table 1 shows the cases, their countries, the year of inscription on the WHL, the starting year of the HUL implementation, and the category of activity. Of the 16 cases, 11 of them feature as part of the WHL, which is 69 percent of the total. Between 1978 and 1990, seven properties were inscribed, while two properties up to 2000 and another two up to today, have been nominated.

These numbers indicate a higher concern for applying HUL in properties inscribed on the WHL. Quito in Ecuador was amongst the first group of properties inscribed on the WHL in 1978, and the initiative of undertaking a HUL application process started with a meeting celebrating the occasion of the 37th anniversary as a World Heritage property (Instituto Metropolitano de Patrimonio 2015). La Plata, meanwhile has tried to be inserted on the WHL since 1997 (Pesci 2017) and according to recent attempts, the notion of HUL has supported the theoretical framework to better understand the OUV in the preeminent planned city awarded with two gold medals at the 1889 Exposition *Universelle* in Paris (Alcorta 1890). Apparently, initiatives start with an inflection point and not only adopt but adapt HUL to current needs and circumstances.

Table 1. Summary of cases in Latin America implementing the recommendation.

Case	Country	Year WHL	Year HUL	Category
La Plata	Argentina		2014	Project
Recife	Brazil	–	2016	Publication
Olinda	Brazil	1982	2007	Publication
Brasilia	Brazil	1987	2015	Publication
Rio de Janeiro	Brazil	2012	2009	Meeting
Valparaiso	Chile	2003	2009	Project
Bogotá	Colombia	–	2010	Project
Ambalema	Colombia	–	2017	Publication*
Old Havana	Cuba	1982	2013	Publication
Quito	Ecuador	1978	2013	Meeting
Cuenca	Ecuador	1999	2012	Project
Old Guatemala	Guatemala	1979	2013	Publication
Tegucigalpa	Honduras	–	2014	Publication
Mexico city	Mexico	1987	2014	Project
Puebla	Mexico	1987	2014	Project
Querétaro	Mexico	1996	2014	Project

* Publication to be released later in 2017.

The year of application denotes this inflection point, where activities start in every case. More than half of the cases (9) started between 2012 and 2014, while only four and three were initiated in an earlier and later period, respectively. The impact of the 2011 Recommendation on HUL is clearly perceived, whereas a gradual increase remains questionable.

4 COMPARATIVE ANALYSES

4.1 *Cuenca in Ecuador*

Cuenca in Ecuador is one of the cases that adopted the *Recommendation* and where updated information is available. Located in a valley of the Ecuadorian Andes Mountains, Cuenca is the third city in importance in Ecuador after Quito and Guayaquil, the capital and main port city, respectively.

The historic city centre of '*Santa Ana de los Rios de Cuenca*' was inscribed on the WHL during the 23rd Session of the World Heritage Committee in Marrakesh. The city, formerly committed for heritage protection in the 1970's, achieved a national heritage declaration in 1982. A Special Plan for heritage management along with the institutionalisation to protect the areas inscribed in the WHL was also sanctioned in this year. The institutional system has been maintained while an update on the Special Plan has not been possible until today (Siguencia Avila and Rey Pérez 2016).

Academic initiatives have also supported governmental efforts. The vlirCPM research project, in the framework of the Interuniversity Cooperation between Universidad de Cuenca in Ecuador and KU Leuven in Belgium, has been engaged to analyse the HUL approach since 2012. This study was established with the following aims: to broaden the concepts of heritage and management in the city, to analyse the links of heritage and new urban development on the peripheries of the historic areas, and, to involve an active civic participation. Therefore, the principles of the Recommendation on HUL were pursued through literature review and the integration of stakeholders in the research process.

The UNESCO six-step action plan was applied in Cuenca using available data and collecting new resources. This plan contains the following six steps: mapping resources, reaching consensus, assessing vulnerability, integrating development frameworks, prioritising actions and establishing partnerships. All steps were attempted to be fulfilled by the academyled joint project between the vlirCPM, a University project, and an international network cooperation with the World Heritage Institute of Training and Research for the Asia and the Pacific Region (WHITRAP). The government was integrated as a key actor through agreements, but unfortunately, its commitments have not been completed yet (Rey Pérez *et al.* 2017).

Resources were mapped using both Government and non-governmental sources comprised of consultants and residents. The former provided access to its most recent 2013 inventory, housed by the Direction of Heritage Historic Areas (DAHP) and the latter provided reports and workshops. This step was consolidated in a Geographic Information System (GIS) database with the integration of multidisciplinary inputs. After a training process and agreements on the common objectives, fieldwork in the different disciplines was carried out by experts and research assistants. This data includes: population density, land use, heritage value assessment, state of conservation, natural, economic, historic, archaeological and anthropological resources (PUH_C 2015).

The following two steps, reaching of consensus and vulnerability assessment, as well as the fifth step for prioritisation of actions, and were drawn during a workshop with local stakeholders sustained in sociology models. The outcomes reveal that Cuenca, rather than having only valuable inventoried buildings, is a place where most significant components deal with nature, cityscapes, and intangible traditional manifestations. The fourth step, integration of previous data and results, was assembled into series of file records. They contain a perspective coming from the three groups involved in the process in the case of Cuenca: governance, experts, and community. As seen, the steps were not necessarily followed in the same order as in the proposed action plan. For example, that the last step -establishment of partnerships- was the starting point to implement the Recommendation in the case of Cuenca (Siguencia Avila *et al.* 2016).

4.2 *Comparison on the HUL application process*

The seven cases of the application of HUL falling into the category of projects are detailed in this section. An evaluation based on the six-step action plan as well as the main tools used is carried out. Among the four groups of tools, knowledge is reflected in the methods used in projects to implement HUL since most initiatives start from Academia, and to allow that, financial resources are invested. Despite the increasing awareness of governments on cultural heritage (Tweed and Sutherland 2007), regulation tools are still not perceived to be inserted on the HUL implementation process (Sandholz 2016). However, civic engagement tools, according to the case in Cuenca, have proven to be a catalyst for the implementation. For this reason, the comparative analysis depicts the steps done in the frame of the six-step action plan, and based on the three main stakeholders engaged

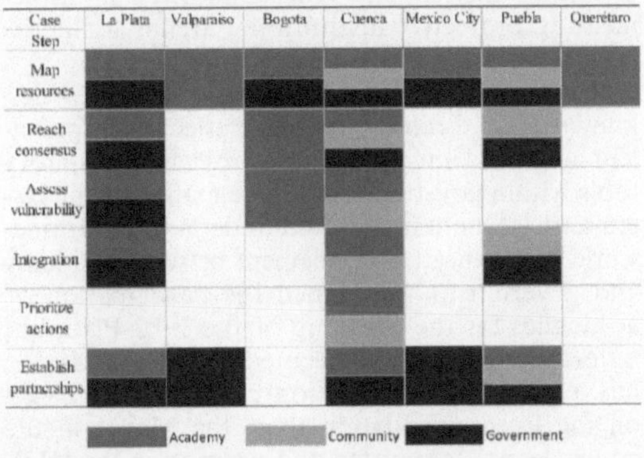

Figure 5. Projects in Latin America and the implementation of the six-step action plan.

in the case of Cuenca: academy, community, and government. Figure 5 shows the seven projects selected for analysis in the mentioned framework.

Within the six-step action plan, the first step—mapping of resources—is normally undertaken in all cases either by academia or governmental institutions. Cuenca and Puebla have involved the community in this process. The data contained in these mapping processes expand beyond the architectural approach. For example, in Mexico City, historic poetry is a relevant source for the narrative of the old city. The next two steps dealing with reaching consensus and assessing vulnerability, are partially developed in half of the projects by academy and government. Integration, as the fourth step, is barely reached in three cases while prioritisation of actions, the fifth step, was only envisioned in Cuenca. Despite this achievement, actions have still not been implemented in the city because of the lack of continuation of the project. The last step dealing with the establishment of partnerships, was assumed by three cases and merely integrated the community in the case of Puebla.

In this analysis framework, the case of Cuenca recognises academy, government and community as the three groups of stakeholders involved in the implementation process. However, the analysis shows how this model overlooks additional groups detected in other cases. The case of La Plata in Argentina is relevant, since reaching consensus targeted another group of stakeholders, the private sector, specifically, high rise building investors (Pesci 2017). In addition, in this case, NGOs and private investors are crucial for the ongoing project. The order of steps, as seen in Cuenca, is also not followed in all projects and, although all of them map resources, many start by launching partnerships, or prioritising actions depending on the current circumstances. Only two projects take place on non-WH cities, not to mention that La Plata aims to reach such recognition. Evidently, in Latin America, cities are committed to implementing protective institutional arenas for heritage preservation at the time of nomination.

When looking at the tools that HUL provides, knowledge is broadly applied, reflecting a prevalent academic interest in most projects. Regulatory tools within governance heritage management systems, are still not introduced under a broad framework. In Cuenca, the agreement between academy and government, envisioned the participation of academics for the updating of the 1982 Plan, but unfortunately was not executed. Puebla in Mexico has a potential introduction of regulation based on the Recommendation since the Mayor in the city is deeply committed and promoting the HUL notion implementation. Financial tools have been provided to initiate some of the projects with a focus on diagnosis of the state-of-the-art and the understanding around the HUL approach in their cities. In Cuenca, even though financial resources were collected with research purposes through partnerships, using private capital or business, for instance, was not proposed as an option. Civic engagement tools can be found mostly in Puebla and Cuenca, where a focus on participation starts at early phases of the project, while in La Plata the goal is first to consolidate business and government actors to collect data from the community in a later stage (Pesci 2017).

In Cuenca, despite the deployment of academic efforts, the lack of definition of actors' roles causes tasks to be performed multiple times. The government has not created spaces for interaction between private institutions, academics and owners for example. This link could help to better understand and follow the roles of stakeholders. In addition, from a financial point of view, private investment brings a potential group of 'business' actors, already considered in the case of La Plata and crucial when dealing with development in historic cities. The business innovation theory includes the Quadruple Helix model providing an analytical framework to understand the role of key actors in a territorial system which comprises universities, industries, government, and civil society (Galvão *et al.* 2017). Helices approaches can therefore integrate different groups of actors and visualise their interaction. According to Yawson (2009), rather than fostering direct contribution by universities for project implementation, interaction gives opportunities for knowledge created at universities to be transferred and exploited by industry and civil society (Yawson 2009).

The insertion of a quadruple helix for stakeholders in Cuenca can improve the implementation of HUL by considering the key contributions from the cases in Latin America. At the same time, cases in Latin America could better share their practices in a common platform. To do so, groups such as the Global Observatory for Historic Urban Landscapes (GO-HUL 2017) provide potential platforms for sharing experiences and results.

5 CONCLUSIONS

HUL suggests using tools to improve heritage conservation, especially in cities of high cultural significance facing rapid urban development. Latin America is the second region with the highest number of cities included on the WHL. This fact, together with the increasing update on heritage values and expertise involved, encourages actions towards urban heritage conservation. In contrast to previous approaches in this region,

HUL provides tools to consider a broader territory for study as well as a broader scope of disciplines involved.

The case studies reflect diverse methods to approach and understand HUL, starting mainly in academic grounds. In the research projects, the six-step action plan is not strictly followed, while regulatory tools are rarely applied. This trend indicates few governmental initiatives and the lack of involvement with other stakeholders. Cuenca in Ecuador demonstrates the implementation of HUL led by academics but still lacking a clear definition of stakeholders' roles. The process developed through stakeholder integration considers various disciplines and sources. Multidisciplinary contributions have informed the urban-architectural vision, while illustrating cumulative effects of research on conserving city heritage. These contributions pooled in community joint workshops, provided material to follow the six-step action plan. The collective process reveals the need to integrate multidisciplinary approaches to current governmental protection systems and not only to remain as a research outcome. Tools and steps followed during the implementation of HUL shift depending on initiatives promoting practical actions. However, measuring their effectiveness is necessary to establish further criteria of feasibility and success of actions under clear roles of stakeholder involvement. To do so, theories of innovative business are available to support participants' involvement procedures.

REFERENCES

Alcorta, S. (1890). La República Argentina en la Exposición Universal de París de 1889: Colección de informes reunidos. P. Mouillot.

Bandarin, F., Hosagrahar, J. & Albernaz, F.S. (2011). Why development needs culture. *Journal of Cultural Heritage Management and Sustainable Development*, 1 (1), 15–25.

Bandarin, F. & Van Oers R., (2012). *The Historic Urban Landscape: Managing Heritage in an Urban Century.* John Wiley & Sons.

Conti, A. (2012). Paisajes históricos urbanos: nuevos paradigmas en conservación urbana. 12.

Conti, A. (2016). Paisajes Históricos Urbanos: nuevos paradigmas en conservación urbana. 'Patrimonio': economía cultural y educación para la paz (*MEC-EDUPAZ*), 2 (10 (5)).

Galvão, A.R., Mascarenhas, C., Rodrigues, R., Marques, C.S. & Leal, C.T. (2017). A quadruple helix model of entrepreneurship, innovation and stages of economic development. Review of International Business and Strategy, 27 (2).

Geddes, P. (1949). *Cities in evolution*. Williams and norgate ltd.

GO-HUL [online] 2017. GO-HUL. Available from: https://go-hul.com/ [Accessed 15 May 2017].

González Biffis, A. (2013). Los centros históricos latinoamericanos: estrategias de intervención, renovación y gestión. Periodo: 1980–2010. Presented at the V Seminario Internacional de Investigación en Urbanismo, Barcelona-Buenos Aires, junio 2013, Departament d'Urbanisme i Ordenació del Territori. Universitat Politècnica de Catalunya, 636–648.

Instituto Metropolitano de Patrimonio, (2015). Encuentro Internacional. "El Paisaje Urbano Histórico como Herramienta del Desarrollo Urbano Sostenible" [online]. Available from: http://www.patrimonio.quito.gob.ec/index.php/inicio/relaciones-internacionales/posicionamiento-estrategico/eventos-internacionales/469-encuentro-internacional-el-paisaje-urbano-historico-como-herramienta-del-desarrollo-urbano-sostenible [Accessed 5 Jun 2017].

Jokilehto, J. (2010). Reflection on Historic Urban Landscapes as a tool for Conservation. *WHP MHC*, 53–63.

Lalana Soto, J.L. (2011). El Paisaje Urbano Histórico: modas, paradigmas y olvidos. *Revista Ciudades*, 14, 15–38.

Lalana Soto, J.L. & Santos Ganges, L. (2011). El problema de las vistas relevantes en la conservacion del patrimonio urbano. Presented at the VIII Congreso Iberico de Urbanismo, COVILHÃ-PORTUGAL.

Lemaire, R. (1915). La reconstruction de Louvain.

Montes Lira, P.F. (2001). El ordenamiento territorial como opción de políticas urbanas y regionales en América Latina y el Caribe. *CEPAL*.

O'Donnell, P., ICOMOS & IFLA. (2012). The Historic Urban Landscape Recommendation: A New UNESCO Tool for a Sustainable Future.

Pesci, L. (2017). La Plata, the urban project of the CEPA foundation. Skype.

PUH_C. (2015). Reassessment of the cultural and natural heritage of the city of Cuenca, based on the sustainable development strategies backed by the HUL approach. Cuenca: Universidad de Cuenca.

Rey Pérez, J., Astudillo Cordero, S., Siguencia, M.E., Forero, J., and Auquilla, S. (2017). Paisaje urbano histórico. La aplicación de la recomendación sobre el paisaje urbano histórico (PUH) en Cuenca (Ecuador). Una nueva aproximación al patrimonio cultural y natural.

Rojas, E. (1999). Old cities, new assets: preserving Latin America's urban heritage. Inter-American development bank.

Rojas, E. (2002). Urban heritage conservation in Latin America and the Caribbean: a task for all social actors. Inter-American Development Bank.

Rojas, E & de Moura Castro, C. (1999). Lending for urban heritage conservation: issues and opportunities. Inter-American Development Bank, Sustainable Development Department.

Sandholz, S. (2016). Urban Centres in Asia and Latin America: Heritage and Identities in Changing Urban Landscapes. *Springer*.

Siguencia Avila, M.E., Pereira Roders, A. & Van Balen, K. (2016). How can the HUL approach improve heritage management in Cuenca? In: Inheriting the city: Advancing Understandings of Urban Heritage, Conference Proceedings. 1–17.

Siguencia Avila, M.E. & Rey Pérez, J. (2016). Heritage values protection, from the monument to the urban dimension. Case study: the historic centre of Santa

Ana de los Ríos de Cuenca, Ecuador. *The Historic Environment: Policy & Practice*, 1–13.
Steinberg, F. (1996). Conservation and rehabilitation of urban heritage in developing countries. Habitat International, 20 (3), 463–475.
Steinberg, F. (2005). Strategic urban planning in Latin America: Experiences of building and managing the future. *Habitat International*, 29 (1), 69–93.
Tweed, C. & Sutherland, M. (2007). Built cultural heritage and sustainable urban development. *Landscape and urban planning*, 83 (1), 62–69.
UNESCO. (2005). Vienna Memorandum on "World Heritage and Contemporary Architecture – Managing the Histor ic Urban Landscape".
UNESCO. (2011). Recommendation on the Historic Urban Landscape. Paris: UNESCO, Records of the General Conference No. 36 C/Resolution 15.
UNESCO. (2017). World Heritage List [online]. World Heritage List by region. Available from: http://whc.unesco.org/en/list/&order = region.
UN-Habitat. (2013). Annual Report 2012. Nairobi: UN-Habitat.
Vandesande, A. (2017). Preventive Conservation Strategy for built heritage aimed at sustainable management and local development. PhD dissertation. KU Leuven.
Veldpaus, L. & Pereira Roders, A. (2013a). Urban Heritage: Putting the Past into the Future. *The Historic Environment: Policy & Practice*, 4 (1), 3–18.
Veldpaus, L. & Pereira Roders, A. (2013b). Historic Urban Landscapes: An Assessment Framework.
Verbruggen, R., Roders, A.P., Stash, N., Leony, D. & de Bra, P. (2014). Protected Urban Planet: monitoring the evolution of protected urban areas worldwide. studies, 7 (4), 319–338.
Whitehand, J.W. (2007). Conzenian urban morphology and urban landscapes. In: Proceedings of the 6th International Space Syntax Symposium.
Yawson, R.M. (2009). The ecological system of innovation: A new architectural framework for a functional evidence-based platform for science and innovation policy.

Innovative Built Heritage Models – Van Balen & Vandesande (Eds)
© 2018 Taylor & Francis Group, London, ISBN 978-1-138-49861-7

Public private partnership and change management for built cultural heritage

C. Boniotti
Department of Architecture, Built Environment and Construction Engineering, Politecnico di Milano, Milano, Italy

ABSTRACT: Considering the deficiency of financing which public bodies are expected to give as an economic support to the management of their own asset, the preservation process of built cultural heritage requires strong alliances between multiple actors from public, private and non-governmental sectors, directed at ensuring conservation and valorization activities upon public built heritage.

Since recent viewpoints strongly promote models aimed at converging investments from different productive sectors into the cultural heritage field and also the European directives are currently encouraging the participation of private resources in valorization interventions, a new financial instrument could be represented by the adoption of Public Private Partnership (PPP) schemes. The potential of each PPP operation requires a re-examination of public policies and administration. It changes the past perspective and forces legislators and public decision makers to put into discussion their profiles, activities, balance of power, the definition of public service and the way to guarantee it.

1 FUNDING STRATEGIES FOR PUBLIC BUILT CULTURAL HERITAGE

The object of the analysis is represented by the public built cultural heritage, owned by the different levels of government based across each national territory. Public properties constitute a considerable inheritance with a strategic value, as well as a social and economic potential, that frequently suffer a condition of neglect and decay. Considering that they represent an asset whose existence must be guaranteed by the State, that their management produces high fixed costs compared to the variable revenues derived from their fruition and the current deficiency of public financing, an analysis of the application of possible business models providing subsidizations is needed. The aim is to find new opportunities to finance the public buildings' management process, assuming that it is comprised of both conservation and valorization activities and that valorization does not mean exploiting cultural heritage from a financial perspective, but rather protecting it and giving it back to the community through its conservation. Besides, the demand for public resources has to be considered not only in terms of money, but also in terms of skills (Macdonald, 2011, p. 893).

A recent model fosters the introduction of non-heritage funding derived from the revenues of other sectors into the culture heritage field, in order to reach heritage and non-heritage objectives. Whilst previous theories stated that the investments directed towards cultural heritage create positive impacts in the economic, social and environmental domains, recent viewpoints claim that merging investments from these different productive sectors into the cultural area may be an opportunity to intercept new financial resources. This approach implies a shift from a "downstream perspective" to an "upstream perspective", which maximizes the cultural heritage impact enhancing it with resources from other domains, such as labor market, regional development, creative industries, digital technologies, etc. (CHCfE, 2015, pp. 195, 196). The possible participation of private resources and the interaction with different productive sectors may represent a contribution to intercept financing originally not directed to heritage.

Cooperation among the different stakeholders is essential to develop these kinds of strategies, where diversified investments interact with the cultural field. The exchange of benefits requires negotiation dialogues directed at sharing good practices and abilities, the involvement of new kinds of public and the creation of networks. It implies the constitution of a trading zone, which is a dialogue between the different parts, despite the divergences in languages, methods and cultural systems (Gustafsson, 2011, pp. 24–26, 56–57, 100, Balducci & Mäntysalo, 2013, p. 2).

Furthermore, the Horizon 2020 Expert Group on Cultural Heritage, within an initiative promoted by the Directorate-General for research and innovation of the European Commission, highlighted that the public sector should consider the participation of private resources in valorization interventions and should refocus its own approach, incentivizing and encouraging the private sector to get involved and invest in cultural heritage. The recommended financial instruments are tax breaks, differentiated VAT rates, well designed grants, loan programs and public private partnership schemes (Council of the European Union, 2014, European Commission, 2015, p. 8).

These positions were already strongly promoted in mid 2000 by the Council of Europe in the report "The private management of cultural property" of the Parliamentary Assembly held on 24 October 2005 (Parliamentary Assembly of the Council of Europe, 2005). Besides, the European Union's Seventh Framework Programme for Research and Technological Development has recently supported the project RICHES, that elaborated guidelines for heritage public private partnership (Promoter srl, 2016).

2 PUBLIC PRIVATE PARTNERSHIP

Once the cultural heritage field is regarded as a positive economic driver and it demonstrates its propensity to collaborate with other productive sectors, it is inevitable for the private to become more deeply involved in exploiting and optimizing its potential at large through its own business model.

Since these recent viewpoints strongly promote programs aimed at converging investments from different productive sectors into the cultural heritage field and also the European directives are currently encouraging the participation of private resources, a new financial instrument could be constituted by the adoption of Public Private Partnership (PPP) schemes.

It can represent a trading zone and an alternative financing mechanism for the conservation and valorization of public built heritage. The interest in partnerships is related to the fact that they are an organizational issue implying a cooperation between different organisms and a multiplicity of interactions. The collaboration developed consists in combining different competences derived from the various sectors. It is the expression of the need of integration between the different public bodies, private companies, non-profit entities, etc. It represents the necessity to link together activities, functions, experts, roles, in order to face the complexity of a project (Codecasa, 2010, p. 175).

2.1 *Definition and characteristics*

The adoption of the PPP could constitute a new financing model entailing a long term collaboration between public and private entities, directed at the realization of public duties (design, construction, management, maintenance of public works or public services), in which resources and risks are proportionally shared on the basis of their own skills (Cori & Paradisi, 2011, p. 41). It identifies a transaction between subjects embodying an idea of public, that can be called "governmental actors", and subjects embodying an idea of private, that can be called "non-governmental actors", aimed at the production of goods or services for the community (Codecasa & Di Piazza, 2010, p. 3; Codecasa, 2010, p. 151).

Even if the PPP definition is not yet completely delimited to the diversified financing arrangements through which it is applied, some features can be considered effective and constant. One aspect is financial pooling, in order to face situations of complexity, uncertainty and risk. In fact, risk analysis has an important role in PPP schemes for the achievement of the value for money. Besides, knowledge and responsibility pooling are also central, as a form of co-production between subjects who pursue different logics (Gualini, 2010, p. 113).

Although the definition also depends on the country, all the PPPs imply the sharing of the three Rs: resources, responsibilities and risks (Macdonald & Cheong, 2014, p. 15).

2.2 *Opportunities for cultural heritage field*

The gathering of resources of a differing nature and the combination of different cultures and interests expressed by partners involved create synergies, constitute an added value and a potential and define innovative policies and solutions. The sharing of ideas, in addition to the sharing of financial resources, fosters the involvement of weaker partners, that are usually marginalized in situations focused on economic aspects (Gualini, 2010, pp. 113, 115, 116).

The main advantages can be considered:

- The mix of public interest, represented by the social utility, and private interest, represented by the profitability.
- The efficiency, in reference to the time and to the ways of implementing the project as necessary conditions for the achievement of expected cash flows.
- The effectiveness, as a public activity transformed into a business is evaluated through profit and competitiveness criteria.
- The transparency, as the involvement of subjects required to work in close interdependence provides cross-checks (Cori & Paradisi, 2011, pp. 43, 45).

In the cultural heritage field, PPP may represent a new business model where private partners bring their financial capital and expertise and are remunerated through proportional risk sharing of revenues generated.

A high cultural cooperation network and a high private patronage are the expression of the mobilization of social and economic actors in the governmental process, primarily on a local scale (Capello, 2016). Cooperation is key in the search for innovation at a regional level, where different subcultures interact to develop a local coordination between theories and actions and to share conceptual and physical instruments aimed at giving shape to a project (Mäntysalo et al., 2013, pp. 11, 13). The complexity of the networks established, which is the indicator of the diversity of the actors involved, and the density, which represents the closeness and the existence of direct links forged, are the measures used to evaluate the quality of a governance (Dente at al., 2005, pp. 42). Besides, private investments create positive impacts at a social and economic level, bringing beneficial externalities to the local and regional development processes (Murzyn-Kupisz, 2013, pp. 156).

2.3 *Diversified financial arrangements*

PPP can be implemented through several and diversified tools and juridical models, from voluntary agreements to procurement forms. They were introduced in the last few decades in Europe and can vary on the basis of the field they belong to and on the basis of each national legislation. PPP represents a flexible model, which refers to a wide range of management alternatives and instruments, each aimed at different public objectives (Benecchi, Savi, 2010, p. 66). Nevertheless, all these forms of joint participation between public and private subjects are directed to the use of financial, managerial and creative private resources, in order to realize or manage public service activities and facilities (Moroni, 2010, pp. 101, 102).

The definition is currently widely debated and contested to the point that the PPP phenomenon is considered as a language game and it has become a term that is politically useful in many instances (Hodge & Greve, 2010).

According to Lawrence Martin, PPPs have many meanings and various tools can be used to help governments achieve their purpose, in this context the preservation of built cultural heritage. It does not convey the same thing to everyone, however the important aspect is to focus on the financing arrangements and on the specific tool. The PPP types can be: Design-Build (DB), Design-Build-Maintain (DBM), Design-Build-Finance (DBF), Design-Build-Finance-Maintain (DBFM), Design-Build-Finance-Operate (DBFO), Design-Build-Finance-Operate-Maintain (DBFOM), with a progressive increase of private sector responsibility (Martin, 2016, p. 2). In fact, the rise of private sector involvement is proportional to the increase of the private sector risk. In the context of cultural heritage field the "Build" phase is replaced with the "Conserve" phase (Rypkema & Cheong, 2012, pp. 13, 14).

One example of DBFOM application in the cultural heritage context is constituted by the restoration of a historical spa called San Pellegrino Spa, in northern Italy. It is the first step of a wider urban regeneration project defined by a Program Agreement, aimed at generating and stimulating new resources across the whole territory in which the historical building is located. The project resulted from public and private objectives. In fact, the work upon the property has been partially financed by a private subject, a real estate group, after an unthinkable discussion with the public authorities (Lombardy Region, Province of Bergamo, Municipality of San Pellegrino). Specifically, the tool adopted is the concession: the public entity will continue to be the owner of the building and the private tenant will manage the structure for thirty years ensuring conservation and valorization activities upon it. Therefore, the participation of private resources, the interaction with different productive sectors and the collaboration with enterprises contributed to intercept financing originally not directed towards the heritage. In the San Pellegrino case, the cultural heritage field interacted with the local tourism, a real estate group, a spa chain and the local industry of bottled spring water.

2.4 *General considerations of legal aspects regulating PPP*

Regulations related to the PPP are sometimes governed by ordinary laws, such as the public procurement legislation, and sometimes governed by special laws, which foresee the control of each individual intervention developed by patrons and the definition of transaction modalities.

This aspect highlights that a partnership constitutes a complex process, without a univocal definitive solution defining the relationship between private and public subjects. It is necessary to take into consideration that the law often does not specify all the aspects regulating the procedure, but it only provides limits in the organization of the way to face problems.

Therefore, since each PPP scheme needs its own specifications, some general requirements are necessary in order to establish a correct contract, such as the clarity of rules applicable to the definition of the agreement. A deal between public and private bodies is a pactional normative itself and it is required to be managed in a transparent way, clarifying the interests of each part. Means, objectives, rules, resources, instruments, methods, controls, reporting, etc. are required to be defined at the beginning of the process (Petraroia, 2016).

As PPP agreements imply a bargaining between governmental and non-governmental actors, an observance of pacts and a correspondence between the execution of work and the requirements determined by the public administration are strictly necessary. It results in quality and image. Consequently, responsibilities related to the monitoring of work progress have to be clearly defined in the coordination of actions developed by legal entities inside and outside the public administration (Codecasa & Di Piazza, 2010, pp. 17, 18).

Aspects that make the process difficult to be managed are:

- The presence of juridical barriers separating the different organizations, as they refer to their own agendas, authorities and balance sheets.
- The public direction may not be completely recognized during the planning and management phases.
- Each organization gives more priority to its own activity planning rather than to the general core activity planning (Codecasa, 2010, pp. 186, 187).

Some measures which could make the procedure clearer are:

- A list of aspects, that an agreement between public and private subjects should regulate. It would contribute to ensure more transparency, avoiding ambiguity and allowing an easy comprehension and management, in order to guarantee continuity and sustainability of commitments. It could provide legality and mutual reliability between the different stakeholders (Petraroia, 2016).
- The individuation of activities, that a public administration should exercise in order to constantly verify the contents and the outcomes of a project (Codecasa, Di Piazza, 2010, pp. 18, 19).
- The control of public objectives and private subject requirements (entrepreneurs and operators of the sector).

2.5 *General considerations of social aspects at the base of PPP*

Beyond normative assumptions, also cultural assumptions are necessary for the development of public private agreements.

Firstly, the perception of the importance of the visibility, which a PPP could offer. It is strongly linked to the development of a social responsibility by the companies. Indeed, in an entrepreneurial culture, the social responsibility of companies requires recognition by the population. Often this kind of commitment is transferred into a social balance sheet, a sort of social reporting.

Secondly, the mutual convenience, particularly developed in the cases of sponsorship.

Then, the will of participation, which is strongly linked to social identification and cohesion. It is particularly developed in the operation of public underwriting and crowd-funding, an activity not widely adopted in the cultural heritage field, but interesting and fundamental, which needs to be implemented. Finally, the will of creation of focused groups, for example in the cases of voluntary work, public underwriting and online petitions (Petraroia, 2016).

2.6 *Private subject role*

In a PPP the private partner is often associated to a profit subject, who acts for the sake of revenues and owns the resources necessary to realize a project.

According to Pietro Petraroia, diversified private actors play in the heritage game (Petraroia, 2013): natural persons; legal persons, which can have or not have for-profit objectives, such as private universities and ecclesiastical entities; economic operators, which work with predominantly private capital and resources; non-profit organizations of social utility; foundations of various types, such as banking foundations; associations (Petraroia, 2016).

Private subjects could provide expertise, know-how, their managerial, commercial and innovative skills, obtaining an economic return. They are expected to contribute to the transaction with funds, debt negotiation and securing, knowledge and skills in real estate development, responsibility for the design or conservation or valorization or long-term tenancy or management (Rypkema & Cheong, 2012, pp. 9, 10).

2.7 *Public subject role*

Thanks to the PPP the public partner could be able to involve more economic operators in the process, improving knowledge and proficiency in its organization, as an answer to society's request.

The public actors can be represented by: institutions, such as local and territorial organisms; entities with a public legal personality, such as functional autonomies (Chambers of Commerce) and foundations established by law (public universities, local health centers); economic operators which work with a financial capital and with resources, often upon singular, predominantly public interventions (they are required to work with a publicist resource management) (Petraroia, 2016).

According to the different circumstances, the public administration should exercise its function of protection verifying the process through different project documents, contracts, establishment of subsidiary companies, groups and project referents (Codecasa & Di Piazza, 2010, pp. 12, 13). Public subjects are expected to contribute to the transaction with incentives in terms of financing and in other supportive ways, regulatory relief, protection of cultural heritage, securing its ownership, ensuring regeneration to the context, infrastructure and services to the asset (Rypkema & Cheong, 2012, pp. 8, 9).

3 CHANGE MANAGEMENT

These models imply some important changes for each partner and within each organization involved, which have to be managed empowering people, improving their awareness as well as skills.

As illustrated in the trading zone processes, the result of these interactions is a change in the behavior of subjects participating in the project, in order to overcome differences. Each organizational change implicates a transformation in people involved. It is part of the contamination procedure and it implies more consciousness of the participation in a shared action. In a trading zone new approaches have to be developed, in coherence with the new situations, strategies and instruments. When different cultures and purposes are merged and move in the same direction, it is necessary to avoid inefficiencies due to the lack of consensus about finalities and objectives. This represents the occasion to acquire new skills, to bring real added value and influence returns on investments.

Some positions claim that partnership should not be considered the goal or an administrative model to follow, but an object of transition, a compromise and a short period solution. It constitutes the first step of a long term process aimed at improving skills and knowledge inside the public and private organizations. The aim is not represented by the agreement, but by the possibilities of redefining and updating public specializations and competences as an answer to society's requests (Codecasa, 2010, pp. 189, 190, 192, 193).

The implementation of change management techniques may facilitate the development of projects in the heritage field, especially when diverse and innovative partnerships are required.

Some experiences fostered initiatives where the stakeholders were transformed into behaviors producers, in order to increase the corporate social responsibility, namely giving back to the society a higher value than the used resources. In these contexts the communication between the actors involved does not exclude possible conflicts but implies a negotiating management and it is indispensable to build partnerships necessary to develop actions. Behavior based projects are aimed at changing attitudes and relationships so that the contractual relationships becomes a partnership (Pennati & Masper, 2012, pp. 31, 45, 46, 55, 95).

One example of change management which proved to be successful in the public sector was developed by the Demanio Agency, the Italian State Property Office. It manages the public real estate, that is constituted of approximately forty seven thousand assets among buildings and land, for a total value of approximately fifty nine millions euros.

In 2004, due to a change of the Agency legal nature, it became a company acting in a private legal regime and it was subjected to a loss of staff. Consequently, a new strategic orientation with an active support was required, in order to develop the economic and social values of the properties. New objectives were defined, in terms of operating and financial efficiency. The general purpose was that each asset has to develop a social and economic value, necessary to cover the management costs.

The decision was to adopt a long-range Strategic Plan, which sets a course of actions over a three-year period, directed at supporting the Agency to successfully act into the global players arena. The activities planned to reconstruct the entity foresaw a revision of processes and organization, IT investments and the creation of new human capital, with a change in the staff. In four years the Agency almost halved its work-force and significantly increased its productivity.

This example represents a quite unusual experience for the Italian public administration, since very frequently public organizational forms do not follow efficiency criteria, but are strictly bound to institutional criteria. In this case, the pressure on results and the private company status fostered an organizational form strongly coherent with objectives (Traini, Rotundo, 2008).

4 NEW PUBLIC MANAGEMENT CULTURE

The fundamental public task to ensure interest in situations where the market is not able to provide solutions, is at the base of the PPP schemes. Moreover, these alternative systems change the past perspective, in which services were exclusively a public issue, and force legislators and public decision makers to put into discussion the definition of public service and the way to guarantee it.

Each PPP project induces the questioning of public administration's profile, organization, activities and balance of power (Codecasa, 2010, pp. 169). Partnership with private means learning to play different roles: the territorial entity, from being a public subject with administrative competences, becomes a local government for the territorial development. A shift from an authoritative, unilateral role to a proactive position occurs, characterized by functions of control, orientation, changes in the selection and management of staff, redefinition of relationship between political and administrative employees and skills in economic evaluations (Antoniacomi, 2010, pp. 22, 23, 24). Knowledge, networking skills, capabilities in managing contracts and negotiating should be fostered.

Besides the public sector, comprised by both technicians and politicians, should also change the timeframe usually spent for the execution of work, that is often decided on the basis of political mandates (Benecchi & Savi, 2010, pp. 66–70).

At the same time the State should not be considered an enterprise and it should not be managed as a company: since it is the only organization legitimated to practice coercion, it is called to act defining strict rules regulating the process and avoiding being judge and jury simultaneously. In support of public actors, the establishment of a consolidated project leadership is the necessary condition for the creation of convenient processes.

Several positions claim that, due to the existence of previous similar procedures such as the foundations, the activation of PPP is not an innovation, and that, considering practical experiences, it is not a guarantee of a successful operation. Moreover, they state that the proactive role of public subjects in projects and market dynamics should be considered an ordinary feature. Nevertheless, the evidence shows that these practices are still currently adopted with several difficulties (Antoniacomi, 2010, pp. 37).

5 CONCLUSIONS

The potential of PPP requires a re-examination of public policies and administration. Since the public sector should refocus its own approach, incentivizing and encouraging the private sector to get involved and invest in the cultural heritage field, the implementation of change management techniques could facilitate the development of partnership projects.

As stated by Susan Macdonald: "*Most governments face significant challenges in their efforts to conserve and manage their cultural heritage assets and few have the necessary resources (money and skills) required to fully achieve their conservation goals. In many places government has been seen as the primary guardian of the nation's heritage, but increasing pressure to fulfill other public demands, requires community commitment and private engagement in order to help governments retain these assets for future generations*" (Macdonald, 2011, pp. 893). Therefore, in a future perspective, the effort to involve subjects not usually interested has to be consolidated, communicating the important role of culture in the local economy and promoting collaboration among actors not usually used to cooperating (Della Torre, 2015: pp. 100). New governance models have to take into account privates as decision makers and as participants in the management of common goods together with public entities. Considering the budget constraints faced by the States, PPPs may ensure private funding and know how in the public sector. The development of PPPs is part of a general public evolution within the economic context, where the State moves from a role of direct operator to a role of organizer, regulator and controller (Commission of European Communities, 2004, p. 3).

ACKNOWLEDGEMENT

The author conveys her sincere gratitude to Prof. Stefano Della Torre from the Department of Architecture, Built Environment and Construction Engineering of the Politecnico di Milano for supervising the current research activity and to Prof. Caroline Cheong from the Department of History of the University of Central Florida for her invaluable comments and suggestions.

REFERENCES

Antoniacomi, G. (2010). L'amministrazione alla prova. Il caso del compendio ex Michelin di Trento. In Guido Codecasa (ed.), *Governare il partenariato pubblico e privato nei progetti urbani*: 21–40. Santarcangelo di Romagna: Maggioli Editore.

Balducci, A. & Mäntysalo, R. (eds.) (2013). *Urban Planning as a Trading Zone*. Dordrecht: Springer.

Benecchi, S. & Savi, I. 2010. Governare il partenariato. Il caso di Parma. "Parma prova e... sviluppa". In Guido Codecasa (ed.), *Governare il partenariato pubblico e privato nei progetti urbani*: 41–70. Santarcangelo di Romagna: Maggioli Editore.

Capello, R. (2016). *Built cultural heritage as an economic developer*. Lesson held at the Politecnico di Milano, Mantua Campus, part of the PhD Summer School "Trading values for preservation and planning in world heritage cities" by Prof. Stefano Della Torre, 5 September 2016.

CHCfE Consortium. (2015). *Cultural Heritage Counts for Europe. Full report*. Available at: http://blogs.encatc.org/culturalheritagecountsforeurope/outcomes/ (accessed 23 June 2017).

Codecasa, G. (2010). Ripensare I governi urbani. Spunti per un'agenda di ricerca. In Guido Codecasa (ed.),

Governare il partenariato pubblico e privato nei progetti urbani: 151–198. Santarcangelo di Romagna: Maggioli Editore.

Codecasa, G. & Di Piazza, F. (2010). Governare il partenariato pubblico-privato. Strategie di governo e strumenti del management pubblico nei progetti di riqualificazione urbana. In Guido Codecasa (ed.), *Governare il partenariato pubblico e privato nei progetti urbani*: 1–20. Santarcangelo di Romagna: Maggioli Editore.

Commission of European Communities. (2004). *Green Paper on public-private partnerships and community law on public contracts and concession*. Available at: http://eur-lex.europa.eu/legal-content/IT/TXT/?uri = CELEX%3 A52004DC0327 (accessed 15 November 2017).

Cori, R. & Paradisi I. (2011). Una ipotesi di lavoro: l'applicazione ai servizi del sistema dei beni culturali del Project Financing. In Alessandro Leon & Paola Verdinelli De Cesare (eds.), *Qualità dei bandi per l'acquisto di servizi nel sistema dei beni culturali. I servizi del sistema dei beni culturali: come interpretarli, combinarli, innovarli, qualificarli*: 41–52, volume 2. Available at: http://www.dps.tesoro.it/documentazione/uval/Vol2.pdf (accessed 23 June 2017).

Council of the European Union. (2014). *Conclusions on cultural heritage as a strategic resource for a sustainable Europe*. Brussels, 20 May 2014. Available at: http://www.consilium.europa.eu/uedocs/cms_data/docs/pressdata/en/educ/142705.pdf (accessed 23 June 2017).

Della Torre, S. 2015. Shaping Tools for Built Heritage Conservation: from Architectural Design to Program and Management. Learning from Distretti culturali. In Koen Van Balen & Aziliz Vandesande (eds.), *Community involvement in heritage*: 93–101. Antwerpen-Apeldoorn: Garant.

Dente, B., Bobbio, L. & Spada, A. (2005). Government or Governance of Urban Innovation?. *The Planning Review*. 41(162): 41–52.

European Commission, Directorate-General for Research and Innovation. (2015). *Getting cultural heritage to work for Europe. Report of the Horizon 2020 Expert Group on Cultural Heritage*. Available at: https://ec.europa.eu/programmes/horizon2020/en/news/getting-cultural-heritage-work-europe (accessed 23 June 2017).

Gualini, E. (2010). Il progetto come arena contesa. Entitlement, responsabilità e riflessività della partnership in pratiche di governance urbana. In Guido Codecasa (ed.), *Governare il partenariato pubblico e privato nei progetti urbani*: 111–150. Santarcangelo di Romagna: Maggioli Editore.

Gustafsson, C. (2011). *The Halland Model. A Trading Zone for Building Conservation in Concert with Labour Market Policy and the Construction Industry, Aiming at Regional Sustainable Development*. University of Gothenburg.

Hodge, G. & Greve, C. (2010). Public-Private Partnerships: Governance Scheme or Language Game?. *The Australian Journal of Public Administration*, 69(S1): S8-S22.

Macdonald, S. (2011). Leveraging heritage: public-private, and third-sector partnerships for the conservation of the historic urban environment. In *Le patrimoine, moteur de développement. Heritage, a driver of development*, proceedings of the 17th ICOMOS general assembly symposium. Paris: ICOMOS.

Macdonald, S. & Cheong, C. (2014). *The Role of Public-Private Partnership and the Third Sector in Conserving Heritage Buildings, Sites, and Historic Urban Areas*. Los Angeles: The Getty Conservation Institute.

Mäntysalo, R., Balducci, A. & Kangasoja, J.K. (2013). Planning as Agonistic Communication in a Trading Zone: Re-examing Lindblom's Partisan Mutual Adjustement. In Alessandro Balducci & Raine Mäntysalo (eds.), *Urban Planning as a Trading Zone*: 7–21. Dordrecht: Springer.

Martin, L. (2016). Public Procurement Practice. Public-Private Partnership (P3): Facilities and Infrastructure. In *Guidance issued by the National Institute of Governmental Purchasing to state and local government procurement officials*: 1–14. Available at: http://www.ncppp.org/wp-content/uploads/2017/02/Public-Private-Partnership-P3-Facilities-and.pd_.pdf (accessed 23 June 2017).

Moroni, S. (2010). Risorse carenti, discrezionalità auspicabile, impreparazione diffusa? Appunti sugli strumenti partenariali. In Guido Codecasa (ed.), *Governare il partenariato pubblico e privato nei progetti urbani*: 101–110. Santarcangelo di Romagna: Maggioli Editore.

Murzyn-Kupisz, M. 2013. The socio-economic impact of built heritage projects conducted by private investors. *Journal of Cultural Heritage*. 14(2): 156–162.

Parliamentary Assembly of the Council of Europe. (2005). *The private management of culture*. Text adopted by the Standing Committee, acting on behalf of the Assembly. Report of the Committee on Culture, Science and Education. Rapporteur: G. Hooper. Available at: http://assembly.coe.int/nw/xml/XRef/X2H-Xref-ViewHTML.asp?FileID=11062&lang=EN (accessed 23 June 2017).

Pennati, A. & Masper, M. (2012). *Sicuri per mestiere. Una storia non ordinaria di sicurezza in cantiere*. Bologna: il Mulino.

Petraroia, P. (2013). Non esiste un solo "pubblico" e un solo "privato". *Il Giornale delle Fondazioni*. 15 November. Available at: http://www.ilgiornaledellefondazioni.com/content/non-esiste-un-solo-%C2%ABpubblico%C2%BB-e-un-solo-%C2%ABprivato%C2%BB (accessed 23 June 2017).

Petraroia, P. (2016). *Partenariato tra pubblico e privato nella tutela e nella valorizzazione dei beni culturali*. Presentation held at the Istituto Lombardo Accademia di Scienze e Lettere, during the Conference "Individuazione e tutela dei beni culturali. Problemi di etica, diritto ed economia", Milan, 7 April 2016.

Promoter Srl (V. Bachi, A. Fresa, C. Prandoni, T. Masi, E. Debernardi) (2016). *Public-Private Partnership Guidelines for CH*. RICHES Project. Available at: http://resources.riches-project.eu/wp-content/uploads/2016/06/RICHES-D7.3-PPP-Guidelines-for-CH_public.pdf (accessed 23 June 2017).

Rypkema, D. & Cheong, C. (2012). *Public-Private Partnerships and Heritage: A Practitioner's Guide*. Washington: Heritage Strategies International.

Traini, G. & Rotundo, F. (2008). Un caso di change management nella Pubblica Amministrazione: l'Agenzia del Demanio. *Economia dei Servizi*, year 3(2): 297–310.

Innovative Built Heritage Models – Van Balen & Vandesande (Eds)
© 2018 Taylor & Francis Group, London, ISBN 978-1-138-49861-7

The participation of the restoration companies into the Preventive and Planned Conservation (PPC): Strengths and weaknesses of the strategies after ten years monitoring

S. Bossi
ABC Department, Politecnico di Milano, Milano, Italy

ABSTRACT: Since the second half of XX century, the Preventive and Planned Conservation of built cultural heritage (PPC) has developed important theoretical advancements, sharing in the more update strategies of the cultural heritage sector. However, in the Italian framework there is again a need to define role and procedure to start up the innovative process, for all the stakeholders involved: administration, technicians and restoration companies also. Starting from the outcome of a PhD research, this paper investigates the activities of the Italian restoration companies in the framework related to PPC strategy. The goal is to understand the actual tendencies of the market, the innovations, the corrections needed, coordinating the different best practices undertaken in the last year, through the experiences of involved shareholders but also compared to international experiences. In conclusion, the paper submits some suggestions about the future development in this market sector with an operative role of the restoration companies.

1 INTRODUCTION

The Preventive and Planned Conservation (PPC) of built heritage is an innovative procedure, where the innovation could be summarize thought as a step from the restoration as an event to the preservation and risk assessment approach as a long-term process (Della Torre 2010). In order to undertake a PPC strategies the most update theories affirm that first of all we have to innovate the cognitive process, which is a very important aspect for the innovation of stakeholders involved.

In the last years several researches have investigated the procedures and capabilities necessaries to undertake this innovative approach both for the public and private administration (Della Torre & Petraroia 2007), for the owner and the managers (Minosi 2007). The building information system dedicated has been studied also (Benatti et al. 2014). Moreover, about ten years ago, a PhD research investigated the role of restoration companies in the Italian framework (Bossi 2009). The main outcome of this research has been the feasibility study of three innovative operative scenario for the SME: the inspection service, the planned conservation service and the draft of conservation plan. The PhD research developed for each scenario a SWOT analysis (Bossi 2010a) and the operative procedures of which in the following pages summarize the principals exits.

The inspection service (IS) is a multi-year service composed of both the inspection activities and the correlated report (that explains how the activities are done). The principal reference was the international organization: starting from the first experience of Monumenten Wacht in Netherlands till to the last developments of Monumentenwacht Vlaanderen (Čebron Lipovec & Van Balen 2010). It's clear that the skills and the experience of the inspection team (usually composed by one architect and one maintainer) are the main aspects to guaranty the quality of the inspection. The service is thought as a first preventive approach (primary prevention): an inductive process based on the experience and skill of the inspection team (Bossi 2010b). It can be undertaken when there isn't a conservation plan of the building, in order to have a preliminary condition assessment. It can be done also as post-intervention control some years after to the restoration work, to check the quality of the activities made and to take under control the buildings. This inspection service could offer also important strengths such as impartiality, dissemination of awareness and a wider responsibility in all stakeholders involved. However, in the Italian context this model has weaknesses also such as a lack of upgrade training and an absence of financial support. Above all, the separation of inspection and maintenance activities risk to reduce quality of maintenance work and their role as inspection's moment. For an analytic description of the service refer to the SWOT analysis attached (see Figure 1).

Inspection service (IS)

Strengths	Weakness	Opportunities	Treaths
Low and sure cost for the customer	Low capability to plan of the contracting	Flat rate and less inconveniences (not planned activities)	High cost for not planned activities
Possibility to unite with the restoration intervention	Conflict of interest if the company carry out also the maintenance activities	Loyalty marketing	Inapplicable service
Interesting service for real estate of historic building	Additional cost for the customers	Avoid the big damages	Useful not understood and then the service is not undertaken
Inspection and maintenance separated: minimum preventive level	Keep to the material authenticity	More attractive for the customer	Risk to lose the aim of the inspection activity
Advertisement of the company core business	Conflict of interest if the company carry out also the maintenance activities	Increase of customers Spin off possibility	Enterprises competition not based on service quality
Optional service	Need of incentives for companies and customers	The owner shares principles and useful with the company	Service not applicable
Fair service	Not self sufficient	Sharable service	Service not applicable
Service provided for public but also private buildings	Not economic self sufficient without incentives	Dissemination of conservation culture	Few membership amongst not listed buildings customers

Figure 1. SWOT analysis of Inspection Service (IS).

Planned conservation service (PCS)

Strengths	Weakness	Opportunities	Treaths
Innovative service	Service not understood	New market opportunity	Lower demand
Qualified service	Need to qualify the outsorcing assignment	Innovative skills	Service don't understood and then not activated
Surer and longer cashflow for the enterprise	Multi-year contract. Economic sustainability	Innovation in the legal framework and the market	Conflict of interesting
Self sufficient service	Need of incentives for the customer only	Less tax for not listed buildings and funding for listed buildings	Service not applicable
Loyalty marketing Continuity of the technicians in the building field	Underestimate of the technical and professional weakness	Service as an assurance after the restoration intervention Know how as an advantage for enterprise	Underestimate of critical issues of the building
Service more attractive with other facility	Company consortium	Restoration company as a main contractor	Disadvantages for the company leader
Integrated system	Big investment	Use the information system	Any warrantee for service
No global service	Complexity of service	Service more attractive with other facilities (cleaning, gardening,...)	
At the same time inspection and maintenance activities	Double risk: excessive unplanned works; low attention in planned works	Qualification of maintenance; attractiveness of conservation plan	Harm to the material authenticity of the building
Check and management in housing	Need of qualified contracting	Corrective actions	Problems between different stakeholders involved
Constantly managed information during the	Company must have an adequate management tools	Qualified company	Bad information management

Figure 2. SWOT analysis of Planned Conservation Service (PCS).

In the second service, the Planned Conservation Service (PCS), the controls (like the first) are linked to maintenance and restoration activities. It means that the company is involved in the execution and registration of all activities of the conservation plan. This service could be offered both after restoration work and as independent service. In this second way, the designer of the conservation plan can be either "in house" (or "inside") of public/private owner or an external professional. This kind of service can be an assurance for the customers after having complete the restoration work. Furthermore, an important opportunity of this scenario is that the company's financial aspect is satisfied through the incomes related to the maintenance activities. However, the management of the service is complex and if is not adequately monitored, the costs could increase due to unplanned works. For an analytic description of the service refer to the SWOT analysis attached (see Figure 2).

In the last scenario, the Draft of Conservation Plan (DCP), the restoration companies could make the activities but also the draw of the conservation plan. During the works the restoration companies are the main actors in the process for the knowledge of the building. However, it's important to underline that this scenario, started with company need, is very dangerous for the building process due to the double role of the company, both the maker and the designer. The SWOT analysis of the DCP service is not developed because is very hard to have a real and safe market.

The SWOT analysis summarized in the previous paragraph underline that the restoration companies could be play an important role in the innovative scenario, not only in the maintenance activities, but also in the inspection service for many years. Due to the peculiarities of cultural contest, the previous studies were orientated to sustain the integrated service (PCS), where the inspection was both managed and executed with maintenance activity.

2 ACTUAL EXPERIENCES (2007–2016)

After the PhD research, different kind of experiences has been developed in the preventive and maintenance sector. This paper studies the typologies, and the opportunities of this kind of experiences. In a different way, compared to the hypothesis previously analyzed, the integrated service hadn't a good and hoped outcome: the best-practices was not so much, and it was necessary to modify the rules of the current market to undertake these experiences. Moreover, in these years there were several experiences of inspection activities: for instance, the experience of churches with Diocesi of Piacenza or the inspection service for the critical analysis of the current state in Valtellina of forty-three buildings restored with the grant of Law 102/90 (Gasparoli et al. 2014). These experiences demonstrated that many issues are not solved yet: is it true the role of companies is related to inspection activities only? When an inspection activities service is "sustainable"?

Furthermore, in the same period there was also service of conservation plan and inspection activities in the central archeological area of Rome and Ostia. This most important experience started under the order of the Presidency of the Council of Ministers called "implementation of emergency measures necessary to overcome the situation of danger in the archeological sites of Rome and Ostia Antica" (Cecchi & Gasparoli 2012). This experience was similar to MOP (Maintenance cost analysis), the new service started in 2011 by Monumentenwacht Vlaanderen in which the execution of inspection activities are linked to the drawing of conservation plan and related costs (Vandesande & Van Balen 2016).

Another important project near to MOP, started by Cariplo foundation also, is the project called «to a Monumentenwacht model in Lombardy Region». In this project three different administrations have been involved with a plurality of goals: a call for inspection activities for three historical buildings and a management strategy to drive towards preventive activities planned. In this approach is necessary the public owner take over the general governance in order to assure a continuous process of care, monitoring and control of the conservation process and the selection of professional's companies. In the following year, the public call of inspection activities was won and developed by another restoration company: not a building contractor specialized in restoration (general works: OG2) but a single restorer (specialized in historical surface's works: OS2). This aspect is not less important for the quality of the report: the risk is that a single restorer hasn't the skill to analyze the building as a complex system and to do a condition assessment of the whole building. It should be stressed that only the interdisciplinarity of the team inspection could insure the quality of condition reporting system. The second part of the project was focused to create an information system to manage the conservation aspect. Thanks to another important project, the public administration tested Planet built heritage program (Bossi 2013).

Summarizing only the choices of the restoration company involved in the PhD research, we can affirm that in the same period the company has developed, in addition to the core business activities (restoration works), an innovative branch called «preventive and planned conservation», dedicated to inspection activities and planned conservation service.

Moreover, there are some critical issue: first, the company has undertaken only inspection activities financed by bank foundation; there wasn't an independent commission. Second, the innovative activities undertaken regard only pre-intervention inspection. Neither one order of the post-restoration inspection was activated. For the company this kind of service is attractive only if linked with a new order (as an improvement in the application). For them, it's more difficult to propose this service to old customers, for the company which means a declaration of low quality of for their past restoration made.

Overall a big weakness in the company is the lack of communication and valorization of their innovative activities: the company promoted the inspection service only through their website; the post—intervention service was promoted with a simple mail address to their old customers.

Anyway, to understand the possible future development, it's interesting to compare the area of interest between Italian proposes and Monumentenwacht model, referring to Vlaanderen experiences (see Figure 3). Concerning to the graphic, the inspection activities represent the actions shared between the two models. In both model the inspection activities are related to another service, though the two models are a very different nature (profit and no-profit organization).

The MW model in the last ten years has been an important upgrade. For example, the organization of Monumentenwacht Vlaanderen has demonstrated that the inspection service is not sufficient to promote preventive approach: the MW experience teach that in addition to the report is necessary to quantify the maintenance works and then in other words to draw a conservation plan. Only in this way, the owner is stimulated to undertake this preventive approach.

The Italian experience affirms the same, although the starting point is very different: the inspection report is not sufficient. Not only in terms of conservation raison linked to the PPC theory (already discussed in several scientific papers) but also from a point of view of the restoration companies very different compared to the Monumentenwacht model, because the restoration company is a profit organization. This means the company chases the profit and the profit is assured by an integrated service.

Moreover, both experiences affirm that to started this process is necessary a «critical mass» and to promote a networking between stakeholders.

In this way is important to study the "critical level" of these services in order to be "economically sustainable". Related to the analysis of this question, the financial concept of the break event point is useful to analyze the single component of each service. The "Break even" represent the point in which total cost correspond to the total income,

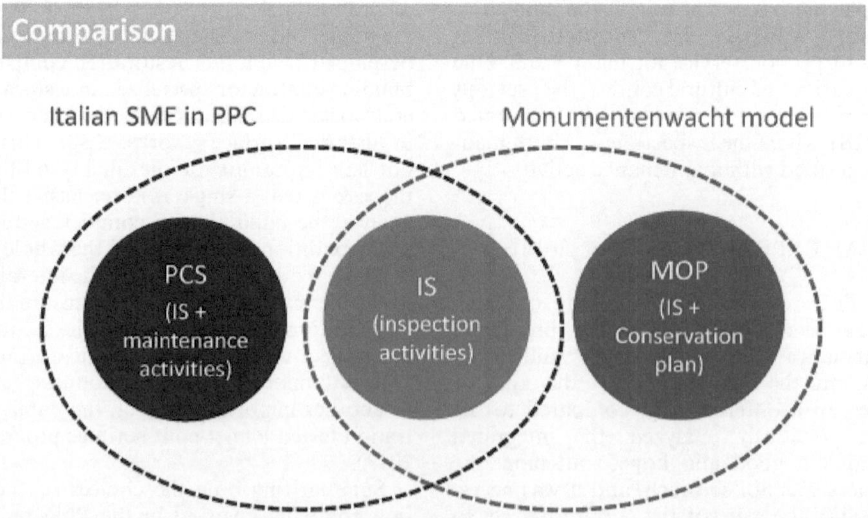

Figure 3. Area of interest of principal services offer in the Italian model and in the Monumentenwacht Vlaanderen model: the in-spection activities represents the actions shared in the two models. It's clear the for both model—though with very different nature (profit and no-profit organization)—is necessary to relate the inspection activities with another service.

it's a financial concept, relevant for a profit company, useful to understand the principal factor that influence the cost of each service. About this kind of analysis, the more quantifiable service with this approach is the inspection service, because it isn't affected by material costs. In this way the research quantified the component of the cost of the inspection service, in order to calculate the break-even point of inspection activities for a restoration company. The main goal was to calculate sufficient numbers of buildings to investigate, necessary to correspond the total cost (see Figure 4).

The analysis of Italian model has demonstrated that the principal factors of the cost of the service are the time (for inspection and for report), related to the localization of the survey (as distant to the location of the organization).

About this analysis there are some important analogies between two models: typologies of costs, the economical aspect, the operative dimension: amount and geographical area of interest. First, the analysis listed similar numbers of buildings for each inspection team: the Italian company estimates 80 buildings (ecclesiastic typology) for one inspection team every year with an average of 30–40 hours for buildings inspection (Bossi 2010b). The Vlaanderen model has a permanent staff of 31 monumentenwacters, who perform 1.30 visual inspection every year, an average of seventy or eighty buildings for one inspection team (Vandesande & Van Balen 2016).

Second, a "local" area that is an efficient dimension for the service: in fact in Italy the provincial area, that is considerate appropriate for the sustainable of the service, is like a Flemish level in Belgium, that is the work area of the monumentenwacters team.

3 ACTUAL TREATHS

Despite the growing research interest and promising examples, in Italy this preventive approach hadn't a good and hoped outcome: the best-practices was not so much, and it was necessary linked to co-financial project able to stress the rules of the current market to undertake these experiences.

It depends from many not solved issues, related to financial, cultural and management aspect. To start up this part of economical market, there is the necessity of economical subsidies (dedicated financial tools), as like happened in the legislation framework for "energy aspect".

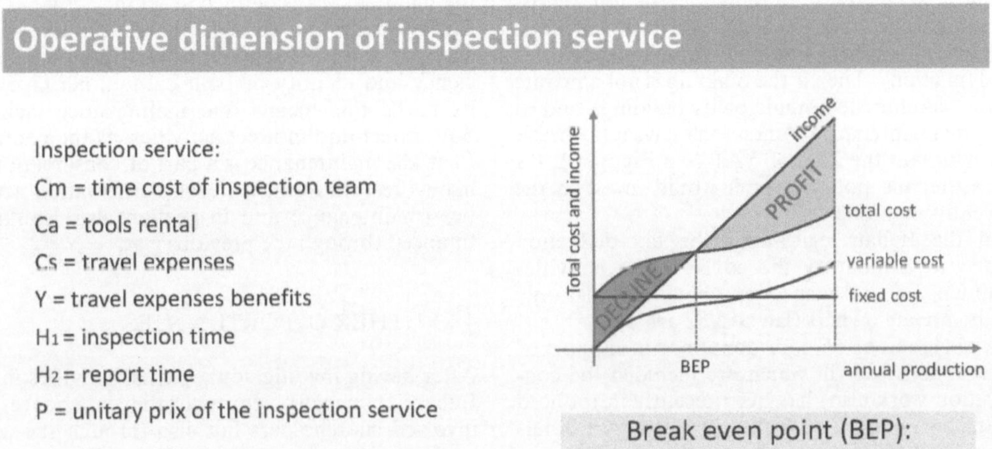

Figure 4. Break-even point applied to the inspection service (IS). The research has quantified the part of the cost of the inspection service, in order to calculate the break-even point for a restoration company. The analysis of Italian model has demonstrated that the principal factor of the cost of the service are the time (time of inspection and time of report), related also to the localization of the survey (as a distance to the location of the organization).

Financial aspects

Restoration works referred to all kind of buildings	VAT 10%
Both extraordinary and ordinary maintenance works referred to "housing private"	VAT 10%
Both extraordinary and ordinary maintenance works not referred to "housing private" (e. g. churchs, oratories, public buildings, historical productive buildings…)	VAT 22%

Figure 5. Different kind of VAT for the conservation activities in Italy.

But if we analyses the Italian legislation we discovery that the tax break (10% VAT inside of 22%) concerns all works of buildings with "housing principal function". Then if the building is not a private house—as churches, municipality buildings, and so on—the maintenance works don't have a tax break and maintain the 22% of VAT (see Figure 5). It's mean that the policy doesn't would invest in the preventive approach.

In the Italian legislation, the tax deduction related to mandatory the conservation activities (maintenance and inspection activities) exist only for the private owners (law 512/82 art. 3).

A tax credit of 65% for donations in support of cultural activities (in which are included the conservation work also) has been recently introduced by Italian ministry of Culture (called "Art bonus L. n. 106/2014").

Finally, in theory, the heritage codex considers a possibility of a direct and indirect contribution by the Ministry (L. Decree 42/2004, art. 35–37), but since 2012 this kind of possibility has been postponed.

Moreover, there are also several mistakes related to the interpretation of PPC activities and their application field, concerning the cultural aspect. If we want to exceed this, it is important to refer to the Heritage code (L. Decree n. 42/2004, art. 29) that affirms there is only one definition of "maintenance": it means the difference between ordinary and special maintenance doesn't exist. Second, the maintenance is defined as all the activities and works carried out to check condition assessment and maintain the integrity, the functional efficiency and identity of built cultural heritage and its parts: this means that maintenance includes both direct and indirect activities, as the controls. Last, the maintenance is a part of conservation as like as restoration, then each conservation activities (maintenance and inspection also) could be financed through the previous way.

4 FUTHER OPPORTUNITIES

After having investigate the actual activities in the Italian framework, through the experiences of involved shareholders but also through the comparison of international experiences, it is correct to indicate the good practices and the actual opportunities to promote this approach.

It's clear that the economic benefits could makes more attractive the preventive sector and in this way, there is a need to promote financial benefits for the restoration companies also. However, the economic aspects are not the only aspects to develop.

In addition to this, the innovation of the sector regards not only the restoration companies, but overall the stakeholders, as the associations of buildings and related companies and the administrations. The innovation of a single restoration company (like as restoration company study in the PhD experience) could not be guarantee a real market. It's need also an innovation of all the sector, starting to the administration proposals.

A careful experience has been recently undertaken by Diocesi di Mantova for the restoration of four historic churches, damaged by earthquakes. Although the ecclesiastical organization are not obligated to make a public procedure for restoration work, the Diocesi of Mantova has studied a public call for conservation works (seismic interventions also), where the quality aspects of the offers were the main parameters to evaluate the proposals.

In this moment, there are three announcements for three historical churches: the announcement ended for the S. Bartolomeo Apostolo church in Quistello (MN) and the S. Giovanni Battista church in Moglia (MN) and an announcement open for the S. Tommaso Apostolo in Bondeno.

In these calls have been integrated a conservation plan designed by ecclesiastic administration with a score proportional to the years offer by the restoration company (from a minimum of two to a maximum of ten years). The first outcomes of this experimentation are very good: all the restoration companies involved had no difficulty to offer this kind of qualified service, instead of an economical discount only. In both cases there are been a reduction of the discount rather than a reduction of quality offer in the PPC strategy (http://www.diocesidimantova.it/it/gare).

The qualitied offer by the restoration companies involved in these calls regards the responsibility to do an annual or biannual feed back to the administration about to the maintenance activities undertaken.

In this way before the final payment, the winner must subscribe with a bank a guaranty to cover the total period offered in the conservation plan. It is a kind of guaranty for the mandatory activities planned in the conservation plan. The bank guaranty will be released each year when the annual works will be executed completely and marked accurately. The missed or partial execution of planned activities imply the deduction of money correspondent to the cost of work to execute during the year.

Quality in the selection process is not the only aspect of this kind of PPC strategy. At the same time, the Diocesi of Mantova investigated and promoted the strategic role of the "monument wacht" of the built cultural heritage, as the users and voluntaries people because they could be a link between the programming and management role of administration and the execution role of the companies. They play a daily control of the activities made by the companies and scheduled in the conservation plan. To reach this goal, it's necessary to train this kind of people. In this way, the Diocesi organized training course and a free and online publication in the official website of Diocesi as practical aid for them (Bossi 2016 free download in http://www.diocesidimantova.it/it/beni-culturali).

Concerning the role of the administration, in order to promote the maintenance market for restoration companies, it is important to remember also a current project started in 2014 and promoted by Diocese of Milano related to the technical dossier of ecclesiastic buildings. Although the project was developed with a technological approach, it was very strategic for dimension and for hypothetical impact in the maintenance market: the project involved about 5.800 buildings of which 2.500 churches (Lazzaroni 2014).

Finally, the role of architects or engineers of each province could be strategic to promote a preventive approach, with the temporary professional groups dedicated to inspection activities addressed to the public administration or private owners. In this way an important action has been carry out in the last years by Cultural District of Monza and Brianza as kind of front-office for the owners to offer the services related to the PPC process (Moioli 2013).

5 CONCLUSION

The paper has underlined the most updated attempts in order to start a PPC market, analysing both the role of the restoration companies and the interventions of different stakeholders involved. Referring to the analysis undertaken, and the case studies developed, the paper has demonstrated that the technical sector is ready for this kind of innovation.

In every experience mentioned before, is clear that the change approach is possible only if each activity is linked to process. Is not sustainable a change only limited to one aspect or to one stakeholders of the process. It's important to understand the change must be undertaken only by a holistic approach, including all different stakeholders, which form the core of economic and cultural assessment of the preventive and planned conservation approach.

REFERENCES

Benatti, E. et al. (2014). Planet beni architettonici. Uno strumento per la conservazione programmata del patrimonio storico-architettonico in Della Torre S. (ed), *ICT per il miglioramento del processo conservativo*, Firenze: Nardini: 13–30.

Bossi, S. (2009). Process innovation in the conservation of built cultural heritage: the role of restoration company, abstract of PhD thesis in *PhD Yearbook*, Milano: 484–485.

Bossi, S. (2010a). The Requisite Evolution of Small and Medium enterprises for partecipation in an Innovative Conservation Process, in Mälkki, M. & Schmidt-Thomé, K. (eds.). *Integrating Aims. Built Heritage in Social and Economic Development*, Espoo: Centre for Urban and Regional Studies Publications: 181–194.

Bossi, S. (2010b). L'ispezione del patrimonio storico-architettonico: strumento operativo dell'impresa per la diffusione della cultura della prevenzione in Biscontin G., Driussi G. (eds), *Pensare la Prevenzione: Manufatti, Usi, Ambienti*, Venezia: Arcadia ricerche, 2010: 543–552.

Bossi, S. (2013). Riflessioni sul servizio ispettivo attivato per un Sistema di ville gentilizie del nord—Milano in "Rivista dell'Istituto per la Storia dell'Arte Lombarda", n. 9: 59–68.

Bossi, S. (2016). *L'esercizio della cura. Indicazioni per l'uso e la conservazione quotidiana del patrimonio architettonico ecclesiastico*, Como.

Čebron Lipovec, N. & Van Balen, K. (2010). Tra prevenzione e manutenzione: I "Monumentenwachten" in Biscontin G., Driussi G. (eds), *Pensare la Prevenzione: Manufatti, Usi, Ambienti*, Venezia: Arcadia ricerche, 2010: 193–202.

Cecchi, R. & Gasparoli, P. 2012. *Preventive and Planned maintenance of protected buildings*, Firenze: Alinea.

Della Torre, S. & Petraroia, P. (2007). Integration of maintenance and management activities with authorization procedures for listed architectural heritage, in *Maintenance Management 2007*, Roma: 87–90.

Della Torre, S. (2010). Economics of Planned Conservation in Mälkki, M. & Schmidt-Thomé, K. (eds.). *Integrating Aims. Built Heritage in Social and Economic Development*, Espoo: Centre for Urban and Regional Studies Publications: 141–155.

Gasparoli, P. et al. (2014). La pratica dei processi di cura nella conservazione degli edifici ecclesiastici, in Della Torre, S. (ed) *Metodi e strumenti per la prevenzione e la manutenzione*, Milano: Nardini: 67–78.

Lazzaroni L., (2014). Conservare, mantenere e valorizzare nella Diocesi di Milano, in Della Torre, S. (ed) *Metodi e strumenti per la prevenzione e la manutenzione*, Milano: Nardini: 161–168.

Minosi, V. (2007). Historical buildings owned by local authority: problems and proposals for conservation, *in Maintenance Management 2007*, Roma: 345–346.

Moioli, R. (2013). *Lo sportello per la conservazione preventive e programmata: processi e progetti per l'attuazione*, in "Rivista dell'Istituto per la Storia dell'Arte Lombarda", n. 9: 29–44.

Vandesande, A. et al. (2014). From a working definition of preventive and planned conservation towards the innovative services of monumentenwacht Vlaanderen, *in Della Torre, S. (ed) Metodi e strumenti per la prevenzione e la manutenzione*, Milano: Nardini: 25–34.

Vandesande, A. & Van Balen, K. (2016). An operational preventive conservation system based on Monumentenwacht model in Van Balen, K. & Versrynge (eds), *Structural Analysis of Historical Constructions—Anamnesis, diagnosis, therapy, controls*: 217–22.

Preventive and planned conservation

Innovative Built Heritage Models – Van Balen & Vandesande (Eds)
© 2018 Taylor & Francis Group, London, ISBN 978-1-138-49861-7

Preventive conservation applied to built heritage: A working definition and influencing factors

A. Vandesande & K. Van Balen
Raymond Lemaire International Centre for Conservation and Department of Civil Engineering, KU Leuven, Heverlee, Belgium

ABSTRACT: Preventive conservation of built heritage is increasingly of interest to academic research and professionals in the sector, with promising examples that demonstrate the viability of this approach. Accordingly, there is ample data and material from different research fields, with scopes and focus. Nevertheless, a general overview or commonly held delineation is not yet developed. The aim of this paper is to develop a working definition for preventive conservation of built heritage. The research consists of an inductive analysis based on evidence originating from existing research within the UNESCO Chair on preventive conservation, monitoring and maintenance of monuments and sites (PRECOM³OS) network, literature on general building maintenance and short descriptive case studies, while using influencing factors of preventive conservation practice and implementation as a proxy. The proposed working definition of preventive conservation starts from condition-based maintenance with a monitoring system, which is the least destructive of all interventions which inevitably occur in the built heritage conservation process. While this preventive conservation approach has several benefits, its practice and implementation are characterised by different influencing factors, i.e. the available knowledge and know-how as well as financial, environmental and social factors.

1 INTRODUCTION

Preventive conservation has a long tradition in the fields of museology and archaeology, respectively as a conceptual approach to rapidly safeguard and document archaeological heritage on a site at risk and a practical risk management approach that assures optimum preservation conditions for movable heritage (Van Balen 2011). However, when considering the scope of built heritage, interventions are not limited to immediate response actions and the surrounding environment cannot be controlled. Research and practices related to preventive conservation approaches dealing specifically with built heritage were triggered during the last 3 decades. The topic became increasingly endorsed within international and European policies, as can be observed in the international soft law overview in relation to preventive conservation which was initiated by Cebron Lipovec and Van Balen (2008). A more in-depth research has been conducted by Pickard (2013) who examined the different resolutions, recommendations and provisions in cultural heritage conventions adopted under the auspices of the Council of Europe. These developments are also reflected in different EU-funded research projects and by international organisations that started implementing a pro-active approach towards monitoring and maintenance, such as the World Monument Fund with the Watch programme and UNESCO with the Reactive Monitoring and Periodic Reporting mechanisms. Within these developments, special notice should made of the 2008 FP7 project SPRECOMAH, or Seminars on PREventive COnservation and Monitoring of the Architectural Heritage. In 2009 this resulted in the establishment of the UNESCO Chair on preventive conservation, monitoring and maintenance of monuments and sites (PRECOM³OS) at the Raymond Lemaire International Centre for Conservation (University of Leuven) in collaboration with Monumentenwacht Vlaanderen and the University of Cuenca in Ecuador. The aim of the collaboration is identifying research and educational activities in the field of preventive conservation, as well as developing new tools and techniques to improve preventive preservation strategies while considering a variety of cultural and social contexts.

As outlined at the PRECOM³OS UNESCO chair conference of 2012 (Van Balen and Vandesande 2013), built heritage preservation approaches are mainly focused on curative conservation and restorative treatments, which do not always result in the removal of causative factors. Once conserved, the historic structure can return to an environment leading to further deterioration, likely requiring future interventions, and establishing a reactive pattern of treatment. In contrast, preventive conservation approaches are more effective in maintaining heritage integrity and more cost-effective within a long-term perspective. Thereby the main benefits of a preventive conservation approach for built heritage are considered the cost-effectiveness for built heritage owner-managers (Forster and Kayan 2009), the quality protection of built heritage, the prolongation of the physical service life of buildings and building parts (Vandesande et al. 2014) and the empowerment of local communities in dealing with heritage (Della Torre 2010) and the exploitation of their services in heritage management.

Parallel to these developments, preventive conservation of built heritage also became of interest to academic research outside the PRECOM³OS network (to name some: Myrbakk and Myrbakk 2001, Lloyd and Staniforth 2000, Abdul-Rashid and Ahmad 2011) and promising examples that demonstrate the viability of this approach can be identified. Accordingly, there is ample data and material from different research fields, with scopes and focus. Nevertheless, a general overview or commonly held delineation is not yet developed. The aim of this research is to develop a working definition for preventive conservation of built heritage and define the factors that influence preventive conservation practice and implementation. The research consists of an inductive analysis based on evidence originating from existing research within the PRECOM³OS UNESCO Chair network, literature on general building maintenance and short descriptive case studies, while using influencing factors of preventive conservation practice and implementation as a proxy. The research process entailed critically interpreting the different data sources, categorising and coding the outcomes into larger thematic patterns and verify the results in a larger context in order to facilitate new theory building on preventive conservation.

2 PREVENTIVE AND PLANNED CONSERVATION: A WORKING DEFINITION

The results the research process are structured according to the preventive conservation framework developed by the PRECOM³OS UNESCO Chair, which starts from 3 classes of preventive conservation for built heritage based on the classification of disease prevention in medicine (Gordon 1983):

- Primary prevention: avoid the causes of the unwanted effect (damage);
- Secondary prevention: condition-assessments and monitoring to enable early detection unwanted effects (damage) and respective symptoms;
- Tertiary prevention: avoiding further spread of the unwanted effect (damage) and the formation of new unwanted (side) effects (damage).

2.1 *Maintenance: Avoid the causes of the unwanted effect*

"The successful preservation of an historic building, complex or city depends on its continued use and the daily care and maintenance" (D'Ayala and Fodde 2008: xv). Maintenance is a widely applied strategy, also in different sectors and disciplines and with a wide variety of definitions, scope and characteristics. Over the past decades, its application has considerably evolved due to an increased maintenance need for a variety of assets and properties with complex designs, new maintenance techniques and changing views on strategies, organisation and expectations (Moubray 1997). These shifts in strategies are reflected in the actual scope and definition of maintenance. The technical committee CEN/TC 319 working on "maintenance" within the European Committee for Standardization (CEN), brings together the contributions and needs of the specific sectors in relation to the more general maintenance discipline. The different CEN/TC 319 subcommittees, including i.a. terminology, risk based inspection framework (RBIF), quality assurance requirements for maintenance operations, maintenance process, maintenance performance and indicators and maintenance of buildings, have developed a series of 'horizontal' standards applicable to different sectors and have links with other European and international committees that deal with issues related to maintenance, such as facility and asset management (Talamo and Bonaomi 2015). The CEN-EN 13306 (2010) standard on terminology defines maintenance as the 'combination of all technical, administrative and managerial actions during the life cycle of an item intended to retain it in, or restore it to, a state in which it can perform the required function'. Furthermore, this European Standard specifies generic terms and subdivisions of maintenance. Specifically, preventive maintenance strategies have a long scientific research tradition and a large applicability in different sectors. They are generally defined as regular maintenance actions at predetermined intervals, or according to prescribed criteria, performed on equipment and infrastructures to reduce probability of failure or degradation an extend their life-span (Lewis 1991, Sullivan et al. 2004).

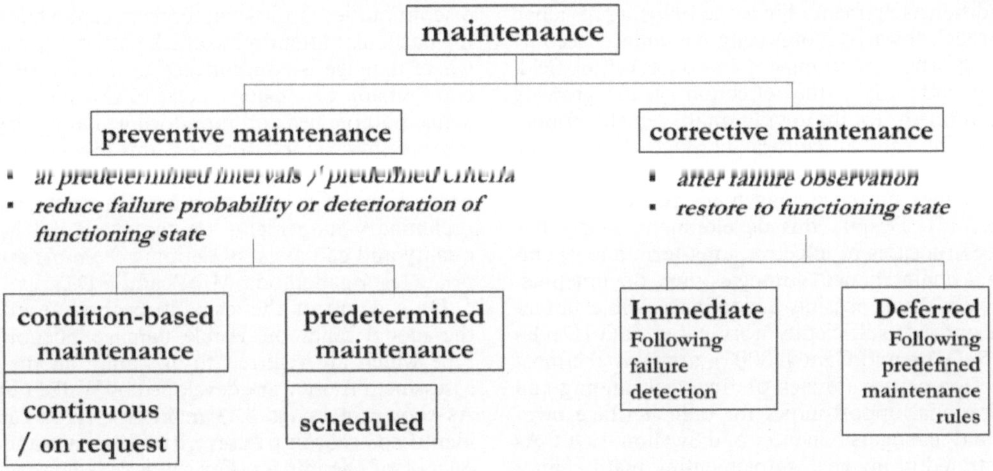

Image 1. Overview of maintenance as defined by CEN – EN 13306 (2010), based on Lind and Muyingo (2012).

Maintenance strategies can vary greatly depending on the context of the maintenance operation (Ahire et al. 2000). For building maintenance, several factors that determine this context e.g. geographical spread, range of environmental exposures, structural complexity, any deterioration in condition and the number of people involved in its management (Bartholomew-Biggs et al. 2006). Choosing appropriate maintenance strategies based on allocation of resources and defined maintenance standards is an important step in the building maintenance operation process (Lee and Scott 2009).

Approaches and arguments for selecting specific strategies are ample (Horner et al. 1997). For each strategy, different management and organisational problems can observed which result from a lack of understanding between maintenance personnel and managing directors who criticise the inefficiency of maintenance implementation (Zavadskas et al. 2010). In the face of decreasing maintenance resources, technology is increasingly a tool to assist personnel and enhance the quality and efficiency of maintenance (Lee and Wang 2008). In practice, the building maintenance strategy choice depends on available resources in combination with common factors, i.a. building characteristics, owner-managers, technical, administrative and political factors (El-Haram and Horner 2002).

Although this overview provides an insight on the possible maintenance strategies for buildings, the two related ideas of standardisation and a common approach are debatable in context of built heritage. Standardisation is applicable to specific technologies, i.a. geometric measurements and Non-Destructive Tests (NDTs), but standardised approaches and strategies to understand and maintain historic structures do not align with the specificities of historic interventions and additions on a building, local building technologies and cultural context. Rather a general rationale can be put forth, i.e. the most effective general maintenance strategy consists of a broad and flexible approach involving economic, social, cultural and technical considerations. Within this general rationale, 'as little as necessary, as much as possible' has become a commonly expressed maxim in regard to maintaining built heritage (Dann and Wood 2004). This observation essentially illustrates that built heritage strategies should not just be based on corrective or preventive maintenance but minimum interventions. Whereas some activities within the scheduled maintenance process are by definition redundant (Horner et al. 1997), regular and periodic condition surveys are identified to be more effective in pinpointing potential defects as early as possible and reducing the number of unnecessary interventions (Miele 2005, Dann and Worthing 2005, Gard'ner 200). Considering that historic structures themselves are the most important knowledge source about historic materials and construction techniques, these minimum interventions are essential to guarantee quality protection within the built heritage sector.

2.2 *Monitoring: Enable early detection unwanted effects*

Achieving an effective maintenance strategy for built heritage necessitates comprehensive data on the nature and condition of buildings (Wordsworth 2001). However, considering that "minimal intervention can only work by utilising regular inspection" (Feilden 2003: 236), this does not imply single condition assessments. Since the 1990s, a pro-active approach towards monitoring has indeed become an important development in conservation field. "This emphasis is the reflection of the growing commitment to improving management frameworks for care of cultural heritage through the use of monitoring, which is understood as a key component of the management process" (Stovel 2008: 15). Despite this development, solely few successful cases of effective, long-term and operational maintenance strategies are documented. Whereas this is possibly due to inaccessible, unreliable and not sufficiently normalised data (Devlin 2005, Dann and Cantell 2007), arguably the most common causes are lack of financial funding and operational opportunities for built heritage owners and managers (Chanter and Swallow 1996). As illustrated in image 2, implementing maintenance strategies based on a regular inspection approach often raises questions on additional costs for setting up a monitoring system.

However, reactive and scheduled maintenance are not cost-effective when measured against the expenditure for a large scale condition based maintenance strategy. More specific, research on monitoring systems on the level of the heritage building stock indicated its effectivity in resource expenditure in contrast to targeting individual historic structures. The most demonstrative example in this context is *Monumentenwacht*, a well-established, experienced and highly professional model that operationalises preventive conservation of built heritage on the level of the heritage building stocks (Vandesande et al. 2016; Vandesande and Van Balen 2016). In practice, the *Monumentenwacht* organisations support private and public built heritage owners and managers in the preservation of their properties through a system of regular systematic maintenance. The *Monumentenwacht* reporting entails pathology, stratigraphic and damage analyses by noting any decay manifestations in relation to building materials and detecting discontinuities in the structural concept while paying particular attention to crack patterns and areas where damage is concentrated as a result of high compression or tensions. Next to revealing (non)-active pathologies and monitoring changes before serious physical deterioration, this model presents the most effective data use for management purposes. It prevents obsolete monitoring data, allows preliminary judgements on the structural health quality and can assist in defining the most appropriate testing methods, MDTs and NDTs.

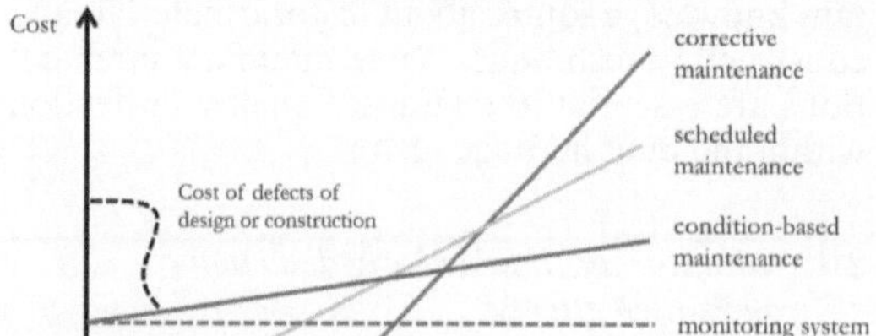

Image 2. Cost relationship between corrective, scheduled and condition-based maintenance with a monitoring system based on Forster and Kayan (2009).

On account of the visual inspection approach, the model relies on visible damage phenomena type which are referred to as "mild" in the risk assessment framework developed by Waller (2003). As shown in image 3, 3 main risk types can be identified according to severity of damage and frequency of the impact. For example, when considering water as deterioration agent, pluvial flooding is a type 1 risk while an environment of high driving rain levels is a type 3 risk.

However, Waller developed this risk categorisation based on a theoretical probability of damage for collections for in a controlled environment. The framework should therefore be reinterpreted for complex historic constructions in an uncontrollable environment. Taking into account the essential link between anamnesis and diagnosis (Taylor 2005, Paolini et al. 2012), the framework is used here to illustrate the link between existing damage phenomena, underlying deterioration agents as well as their relation to future risks. Whereas the model relies on visual inspection, mainly mild damages and their related deterioration agents are assessed during monitoring and in the consequent condition reports. In context of structural and safety assessments, this presents a specific interest as an early warning system for long-term damage accumulations. Without causing major alterations in loading conditions, which are normally expected

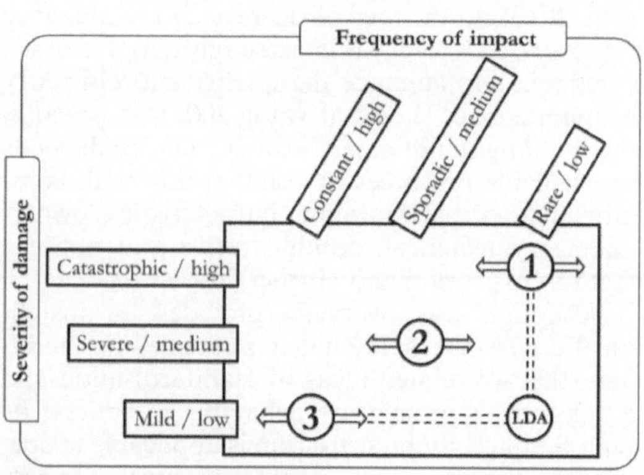

Image 3. Risk assessment framework, based on Waller (2003).

from type 1 risks, these mild damages can also induce historic structure collapses without any warning of the structure being on the edge of its load bearing capacity.

2.3 *Planned interventions: Avoiding further spread of the unwanted effect*

Consequently, preventive conservation starts from condition-based maintenance with a monitoring system, which is the least destructive of all interventions which inevitably occur in the built heritage conservation process. Nevertheless, interventions are necessary to maintain a building's functional, structural and technical performances and extend the life-cycle of built heritage values, authenticity and integrity. In contrast to most corrective and scheduled intervention processes, preventive conservation is based on planned interventions which require a thorough understanding of the building and long-term vision for prioritising interventions. Based on the existing condition of a built heritage structure and the available maintenance resources, appropriate interventions have to be selected and planned. Although the pressing nature of certain interventions can determine the order of implementation, there are no standards or common approaches towards planned interventions. Nevertheless, some key principles and insights can be identified. In his research on the apparent induction period prior to deterioration, Feller (1994) correctly observes that there is a general sense among conservators that deterioration is rarely a linear process but stresses the need for better knowledge and understanding of how complex systems of materials deteriorate. Still today, many practical examples of improper decisions and interventions due to a lack of knowledge and analysis can be observed.

In order to avoid unnecessary detrimental deterioration and specific consequential damage patterns, the most commonly quoted framework in context of preventive conservation is based on the guidelines of the International Council on Monuments and Sites (ICOMOS) International Scientific Committee for the Analysis, Conservation and Structural Restoration of Architectural Heritage (ISCARSAH). The model consists of four cyclical steps: anamnesis, diagnosis, therapy and control, which correspond respectively to the search for significant data and information, identification of damage causes, choice of remedial measures and control of interventions (Heras et al. 2013).

2.4 *The working definition: A practical example*

The final working definition brings together the 3 classes of preventive conservation in 1 approach, which is illustrated here by means of a practical example related to one of the most common problems that affect brick work, i.e. uncontrolled water penetration, for which the causes can vary from windblown rain to inefficient surface drainage and infiltration through failed mortar. Consequently, a variety of biological growths can be found on built heritage structures, i.a. algae, fungi and lichens, for which cleaning the surface of the building is an often opted solution. However, even when implementing cleaning methods which do not increase surface porosity or cause damage to the historic fabric, the result is likely to be ineffective. While cleaning interventions will remove any biological appearance on the surface, it is not certain that the risk is entirely removed. The surface can still be inhabitable for biological colonisation and spores can endure the cleaning, resulting in a probable regrowth or even accelerated growth (Eklund and Young 2013). In order to avoid detrimental damage, a planned intervention starts from a diagnosis of the biological colonisation before implementing cleaning methods. Keeping in mind that cleaning always involve some risks to the historic fabric and that in many cases biological colonisation on brickwork is a natural and harmless process, interventions should only be considered once active damage is identified. Within a preventive conservation approach this is achieved by monitoring the damage pattern. This process can determine whether the growths are in effect causing damage and what their response is to changes in the buildings environment.

In case the biological colonisation is identified as an active damage or the monitoring shows more developed mosses which retain moisture on affected surfaces or even plants with woody roots which can cause pitting and dislodging brickwork,

Image 4. Failure of roof systems and clogged gutters create inhabitable surfaces for biological contaminants (© Monumentenwacht Vlaanderen 2013).

interventions can be planned. However, planned interventions do not start with targeting the damage pattern directly, but from a global view on the historic structure based on adequate building information. The failure of roof systems, ineffective rainwater gutters through a lack of maintenance or damaged pointing are problems which enable the surface to stay inhabitable for biological colonisation. Resolving these problems and drying-out areas of the building that are persistently damp will create a less moist environment and can even cause existing growths to disappear.

3 INFLUENCING FACTORS

Next to the 3 classes of preventive conservation, the research identified different factors that influence preventive conservation practice and implementation. It has been stated that for 'heritage assets, irrespective of both ownership and wider organisational goals, a key specific driver for maintenance should be that it is the fabric itself that is important because of its possible contribution to significance-and not just because of the asset's function' (Worthing and Bond 2008: 189). Next to this top-down expert view, several other influencing factors-and actors-are in place that can be directly or indirectly be associated with preventive conservation. Given that these factors vary for different stakeholders within the built heritage sector and according to local and organisational contexts, solely a generic overview as shown in Table 1 is provided below.

Built heritage professionals and researchers generally acknowledge the synergistic relationship between sustainability and preventive conservation as it entails, more than curative conservation and restoration, preserving heritage values and authenticity, a 'non-renewable capital' (Van Balen and Vandesande 2012). In line with this, a preventive conservation approach essentially prolongs the physical service life of buildings and building components (Thomsen and van der Flier 2009) and therefore indirectly contributes to supporting waste-avoidance (Cassar 2009), preserving embodied energy (Power 2010) and serving low carbon cities (Rypkema 2012). Furthermore, preventive conservation contributes to maintaining the state of conservation and function of built heritage structures. However, these notions should not be interpreted from a condition and resource evaluation perspective. In that context, a roof that leaks a little once every year or solely during strong winds and rain can still be considered in a state which performs the required function. Both from a technical and conservation theory viewpoint, this approach is to narrow for built heritage. At the technical level, buildings are assessed in terms of performance rather than mere function. In their research on quality management of built heritage, Van Roy et al. (2015) define 3 types of performances at 3 levels of detail, i.e. functional performances such as the drainage system, structural performances at the level of the building's structure and technical performances at the applied materials level. Herewith, it is observed that built heritage projects focus mainly on technical and structural performances, causing a lack of maintenance due to decreasing functional performances. Implementing continuous improvements related to use and functional performances can indeed increase the maintenance quality of a building. On the level of conservation theory it should be kept in mind that next to its performances, also the built heritage fabric in itself is an important parameter. Similarly, the distinction between maintenance and repair, or 'retain' and 'restore' as defined by the European maintenance standard, are not interchangeable for built heritage. While repair will increase the buildings performances, also the life-cycle of built heritage values, authenticity and integrity are part of the equation. Next to performances and values, the complexity of built heritage requires even more specific maintenance management strategies. The asset to be maintained entails existing, historical and layered structures that require specific knowledge of historic structures, building materials and specific skills.

Consequently, knowledge takes on an important role within preventive conservation practice. From the state of conservation viewpoint, technical knowledge on building performance, properties of materials and processes of deterioration are pivotal in defining planned interventions. On the other hand, also knowledge on the behaviour

Table 1. Influencing factors for preventive conservation of built heritage.

Knowledge	State of conservation	Financial factors	Environmental factors	Social factors
Maintenance know-how and DIY			Neighbourhood effect	
Awareness of heritage values, authenticity …			Ownership	… and integrity
Behaviour of materials	Building performance	Cost-effectiveness	Life-cycle	Moral obligation
Properties of materials			Depreciation	

of materials is an influencing factor for preventive conservation. Understanding the applicability and durability of a specific intervention in a specific context, as well as know-how on carrying out maintenance and workmanship should be considered into all conservation phases, as those contributions can improve the quality protection of historic structures (Van Balen and Hendrickx 2008). This knowledge is commonly referred to as tacit knowledge or skills (Nelson and Winter 1982) and is based on criteria that were initially decided upon and have been frequently applied with decreasing need for repeated exploration.

From a more practical viewpoint, the owners of built heritage play a very important role in defining the performances for their property and general maintenance behavior. Academic research puts forth age (Reschovsky and Newman 1991), house attachment (Montgomery 1992), type and location of the building (Littlewood and Munro 1996), financial considerations (Yau 2012), available budget (Horner et al. 1997) and gender makeup (Davidson and Leather 2000) which influence owners and managers behaviour towards maintenance. Research dealing specifically with built heritage show that also a considerable amount of owners and managers of historic structures implement maintenance (Dann 2004). However, research on the maintenance behaviour of listed building owner and managers shows that maintenance is mostly not implemented on account of an awareness for heritage values, nor the statutory duty towards their property. The primary motivation is rather 'to avoid the discomfort and financial cost of disrepair and to gain the personal satisfaction of keeping their home in good order' (Hills and Worthing 2006: 203). Furthermore, the research identified social pressure, moral obligation of owners to take care of their property, prior knowledge of house defects and maintenance quality expectations as important factors within the preventive conservation practice. While owner driven maintenance can lead to prioritising esthetical improvements over more preventive conservation oriented intervention, the major concern that can be identified are the effects of maintenance through Do It Yourself (DIY) activities. Whereas DIY maintenance by owners and managers is an important factor to upkeep built heritage, the competence and skills with which it is carried out determine the actual quality of the work. The impact of DYI maintenance increases as it extends beyond mere cosmetic works to basic repair work or even structural improvements. Research on the topic found that the latter is mostly brought about by younger low income owners or by the inability to find competent or trustworthy contractors (Davidson and Leather 2000). This mechanism can result in large consequential damage, buildings that remain in a constant state of disrepair and can deteriorate beyond economic repair.

In this context, the influence long-term cost-effectiveness, considered one of the most important influencing factors in preventive conservation research and practice, should be put into perspective. Although a general lack of knowledge and awareness on long-term savings as well as the direct cost for maintenance interventions is probably a main influencing factor, research also reveals that 'immediate consumption values are the prime consideration for most households in determining repair and improvement priorities' (Munro and Leather 2000: 522). Next to immediate consumption, and in extension place attachment for their property, owners and managers invest to prevent devaluation of their building due to physical deterioration (Harding et al. 2007). The motivation for maintenance investment increases if the owners and managers are more place depended, i.e. if the property is used to conduct their daily economic activities (Graham et al. 2009). Next to these influencing factors in context of built heritage owners and managers, it has been suggested that fiscal incentives or financial premiums can lower the threshold to implement preventive conservation interventions (Maintain our Heritage 2003).

The previous observations mainly approach preventive conservation from a financial viewpoint, which does not adequately capture the complete range of influencing factors. Next to the discussion on individual built heritage owners and managers, it is also relevant to shift the focus to a factor that is often neglected in this context, i.e. the "neighbourhood" effect which deals with the social bonds and interactions within a neighbourhood. Segal (1979: 6) defines neighbourhood as a "physical clustering that provides a collective externality to a well-defined group of people living inside or nearby the cluster". Specifically in context of preventive conservation practice, it should be noted that the neighbourhood effect can also have a tangible influence on the maintenance of buildings. As Galster (1987: 4) states: "Both directly and indirectly, the social neighbourhood can influence the homeowner's upkeep decision. Directly, a strong sense of neighbourhood cohesion can encourage homeowners to maintain or upgrade their property, and can discourage them from falling below neighbourhood norms for minimal housing quality. Indirectly, homeowners' values, perceptions, and expectations, which form the basis of their upkeep calculus, may be altered by the social-interactive dimension of neighbourhood". Since the 1980s, several researchers in social sciences acknowledge the existence of a neighbourhood effect, but it was only more recent that significant (statistic) results on its existence were found (Ioannides 2002). For example, Helms (2012) research on neighbourhood effects in housing renovation. By explicitly modelling the spatial interdependence of households' renovation decisions and analysing a detailed block-level data set, this study finds strong empirical evidence that endogenous neighbourhood effects exists. In this context social bonds within a neighbourhood are of importance, as they enhance the attachment of the inhabitants to the place where they live (Graham et al. 2009). In turn this can have an effect on how communities maintain their neighbourhood or can create longer-term investments in their properties. Next to these environment and social factors, also the awareness for heritage values, authenticity and integrity of communities as well as owners and managers of built heritage play an important role in the motivation to implement preventive conservation.

4 CONCLUSIONS

This article brings together evidence originating from existing research, literature on general building maintenance and short descriptive case studies. In defining a working definition for preventive conservation, the 3 classes of prevention as defined by the PRECOM³OS UNESCO Chair were brought together. The research indicated that preventive conservation approaches differentiate from standards on terminology for maintenance and preventive maintenance strategies for building maintenance. This is largely due to the specificities of historic interventions and additions on a building, local building technologies and cultural context. Rather, regular and periodic condition surveys are identified to be more effective in pinpointing potential defects as early as possible and reducing the number of unnecessary interventions. Thereby, monitoring systems on the level of the heritage building stock indicated its effectivity in resource expenditure in contrast to targeting individual historic structures.

The proposed working definition of preventive conservation starts from condition-based maintenance with a monitoring system, which is the least destructive of all interventions which inevitably occur in the built heritage conservation process. Nevertheless, interventions are necessary to maintain a building's functional, structural and technical performances and extend the life-cycle of built heritage values, authenticity and integrity. In contrast to most corrective and scheduled intervention processes, preventive conservation is based on planned interventions which require a thorough understanding of the building and long-term vision for prioritising interventions.

While this preventive conservation approach has several benefits, its practice and implementation are characterised by different influencing factors. The practice and implementation itself is largely influenced by the need for a monitoring system as well as the available knowledge and know-how as well as financial, environmental and social factors.

In terms of future research, these influencing factors can possibly explain the observation why maintenance management of public built assets gains a significant importance, but the interest of built heritage field in preventive conservation is lagging back. This is especially relevant as environmental and economic concerns are driving investments in the construction sector towards more sustainable approaches.

REFERENCES

Abdul-Rashid, R. & Ahmad, A. (2011). 'The Implementation of Maintenance Works for Historical Buildings—a Review on the Current Scenario', Procedia Engineering, 20: 415–424.

Ahire, S., Greenwood, G., Gupta, A. & Terwilliger, M. (2000). 'Workforce-constrained Preventive Maintenance Scheduling Using Evolution Strategies', Decision Sciences, 31 (4): 833–860.

Bartholomew-Biggs, M., Christianson, B. & Zuo, M. (2006). 'Optimizing Preventive Maintenance Models', Computational Optimization and Applications, 35 (2): 261–279.

Binda, L. & Saisi, A. (2009). 'Knowledge of the building, on site investigation and connected problems'. In: Cosenza, E. (Ed.) Eurocode 8: Perspectives from the Italian Standpoint Workshop, Naples: Doppiavoce Edizioni: 213–224.

Cassar, M. (2009). 'Sustainable heritage: challenges and strategies for the twenty-first century'. APT Bulletin, 40(1): 3–11.

Cebron Lipovec, N. & Van Balen, K. (2008). 'Practices of monitoring and maintenance of architectural heritage in europe: examples of 'monumentenwacht' type of initiatives and their organisational contexts'. In: Kolar, J. (Ed.) CHRESP: Cultural Heritage Research Meets Practice, Lubljana: univerzitetna knjižnica: Kataložni zapis o publikaciji Narodna:238–239.

Chanter, B. & Swallow, P. (1996). Building Maintenance Management, Oxford (Blackwell Science).

D'Ayala, D. & Fodde, E. (2008). Structural Analysis of Historic Constructions. Preserving Safety and Significance, Proceedings VIth international conference, Bath: CRC press.

Dann N. (2004). 'Owners' attitudes to maintenance', Context, 83: 14–6.

Dann, N. & Cantell, T. (2007). 'Maintenance in conservation'. In: Forsyth, M. (Ed.), Understanding Historic Building, Oxford: Blackwell:185–98.

Dann, N. & Wood, S. (2004). 'Tensions and omissions in maintenance management advice for historic buildings', Structural Survey, 22 (3): 138–147.

Dann, N. & Worthing, D. (2005). 'Heritage organisations and condition surveys', Structural Survey, 23 (2): 91–100.

Davidson, M. & Leather, P. (2000). 'Choice or necessity? A review of the role of DIY in tackling housing repair and maintenance', Construction Management and Economics, 18: 747–756.

Devlin, H. (2003). Gutters and VAT, drains and rates. Could Britain learn from continental models of maintenance?, London: Winston Churchill Travelling Fellowship.

Eklund, J. & Young, M. (2013). Biological Growth on Masonry: Identification & Understanding, information for historic building owners, Edinburg: Historic Scotland.

El-Haram, M. & Horner, M. (2002). 'Factors affecting housing maintenance cost', Journal of Quality in Maintenance Engineering, 8 (2): 115–123.

Feilden, B. 2003. Conservation of historic buildings. Oxford: Butterworth-Heinemann.

Feller, R. 1994. 'Aspects of Chemical Research in Conservation: the Deterioration Process'. Journal of the American Institute for Conservation, 33(2): 91–99.

Galster, G. 1987. Homeowners and Neighborhood Reinvestment, Durham: Duke University Press.

Gard'ner, J. (2007). 'Preparing the conservation plan'. In: Forsyth, M. (Ed.) Understanding Historic Building Conservation, Oxford: Blackwell: 156–174.

Gordon, R. (1983). 'An operational classification of disease prevention', Public Health Reports, 98 (2):107–109.

Grahamm H., Mason, R. & Newman, A. (2009). Literature Review: Historic Environment, Sense of Place, and Social Capital, Newcastle: University of Newcastle: International Centre for Cultural and Heritage Studies.

Hardingm P., Rosenthal S. & Sirmans C. (2007). 'Depreciation of housing capital, maintenance, and house price inflation: Estimates for a repeat sales model', Journal of Urban Economics, 61 (2): 193–217.

Helms, A. (2012). 'Keeping up with the Joneses: Neighborhood effects in housing renovation', Regional Science and Urban Economics, 42: 303–313.

Heras, V., Wijffels, A., Cardoso, F., Vandesande, A., Santana Quintero, M., Van Orshoven, J., Steenberghen, T. & Van Balen, K. (2012). 'A value-based monitoring system to support heritage conservation planning', Journal of Cultural Heritage Management and Sustainable Development, 3 (2): 130–147.

Horner, R., El-Haram & M. Munns, A. (1997). 'Building maintenance strategy: a new management approach', Journal of Quality in Maintenance Engineering, 3 (4): 273–280.

Ioannides, Y. (2002). 'Residential neighborhood effects', Regional Science and Urban Economics, 32: 145–165.

Lee, H. & Scott, D. (2009). 'Overview of maintenance strategy, acceptable maintenance standard and resources from a building maintenance operation perspective', Journal of Building Appraisal, 4 (4): 269–278.

Lee, J. & Wang, H. (2008). 'New Technologies for Maintenance'. In: Khairy A., Kobbacy, H., Prabhakar Murthy, D. (Eds.) Complex System Maintenance Handbook, London: Springer: 49–78.

Lewis, D. (1991). 'Turning Rust into Gold: Planned Facility Management', Public Administration Review, 51: 494–503.

Lind, H. & Muyingo, H. (2012). 'Building maintenance strategies: planning under uncertainty', Property Management, 30 (1): 14–28.

Littlewood, A. & Munro, M. (1996). 'Explaining disrepair: Examining owner occupiers' repair and maintenance behaviour', Housing Studies, 77(4): 503–525.

Lloyd, H. & Staniforth, S. (2000). 'Preventive Conservation and 'A madness to Gaze at Trifles': A Sustainable Future for Historic Houses'. In: Roy, A., Perry, S. (Eds.) Tradition and Innovation: Advances in Conservation: Contributions to the Melbourne Congress, Melbourne: IIC and AICCM:118–123.

Maintain our Heritage. (2003). Historic building maintenance – A pilot Inspection service, Bath: Maintain our Heritage.

Miele, C. (2005). 'Conservation plans and the development process', Journal of Architectural Conservation, 22 (2): 23–39.

Montgomery, C. (1992). 'Explaining Home Improvement in the Context of Household Investment in Residential Housing', Journal of Urban Economics, 32 (3): 326–350.

Munro, M. & Leather, P. (2000). 'Nest-building or investing in the future? Owner-occupiers home improvement behavior', Policy & Politics, 28 (4): 511–526.

Myrbakk, G. & Myrbakk, A. (2001). 'Preventive Conservation: For Ever and Ever?', Meddelelser om Konservering, 2: 3–6.

Nelson, R. & Winter, S. (1982). An evolutionary theory of economic change. Massachusetts: Harvard University Press.

Paolini, A. Vafadari, A., Cesaro, G., Santana Quintero, M., Van Balen, K., Vileikis, O. & Fakhoury, L. (2012). Risk Management at Heritage Sites: a case study of the Petra World Heritage Site, Paris: UNESCO and University of Leuven.

Pickard, R. (2013). 'Maintaining the built heritage: a perspective of legal, financial and practical issues derived from Council of Europe actions'. In: Van Balen, K., Vandesande, A. (Eds.) Reflections on Preventive Conservation, Maintenance and Monitoring of Monuments and Sites by the PRECOM³OS UNESCO Chair, Leuven: ACCO: 3–8.

Power, A. (2010). 'Housing and sustainability: demolition or refurbishment?', Urban Design and Planning, 163: 205–216.

Reschovsky, J. & Newman S. (1991). 'Home Upkeep and Housing Quality of Older Homeowners', Journal of Gerontology, 46 (5): 288–297.

Rypkema, D. (2012). 'The Role of Heritage Conservation in a Sustainable Economy'. In: Mälkki, M., Kaisa, S. (Eds.) Integrating aims – built heritage in social economic development, Espoo: Aalto University: Centre for Urban and Regional Studies Publications: 195–211.

Segal, D. (1979). 'Introduction'. In: Segal, D. (Ed.), The Economics of Neighborhood, New York: Academic Press: 3–14.

Stovel, H. (2008). 'Origins and Influence of the Nara Document on Authenticity'. APT Bulletin, 39 (2/3): 9–17.

Sullivan, G., Pugh, R., Melendez, A. & Hunt, W. (2004). Operations and Maintenance Best Practices: A Guide to Achieving Operational Efficiency, US Department of Energy.

Talamo, C. & Bonaomi, M. (2015). Knowledge Management and Information Tools for Building Maintenance and Facility Management, Heidleberg: Springer.

Taylor, J. (2005). 'An integrated approach to risk assessment and condition surveys'. Journal of the American Institute for Conservation, 44: 127–141.

Thomsen, A. & van der Flier, K. (2009). 'Replacement or renovation of dwellings: the relevance of a more sustainable approach'. Building Research & Information, 37(5–6): 649–659.

Van Balen, K. (2011). 'Preventive Conservation in the International context of the PRECOM³OS network', International Conference on Preventive Conservation of Architectural Heritage Proceedings, Nanjing: Southeast University: 1–13.

Van Balen & K. Hendrickx, R. (2008). Preservation of workmanship or workmanship for preservation. In: D'Ayala, D., Fodde, E. (Eds). 2008. Structural Analysis of Historic Construction/Preserving Safety and Significance, Proceedings VIth international conference, Bath: CRC press: 3–13.

Van Balen, K. & Vandesande, A. (2013). Reflections on Preventive Conservation, Maintenance and Monitoring of Monuments and Sites by the PRECOM³OS UNESCO Chair, Leuven: ACCO.

Van Roy, N., Verstrynge, E. & Van Balen, K. (2015). 'Quality management of interventions on historic buildings'. In: Brebbia, C., Hernández, S. (Ed.) Structural Studies, Repairs and Maintenance of Heritage Architecture XIV, Southampton: WIT Press: 313–324.

Vandesande A., Moioli R. & Van Balen K. (2014)." Costing the Built Environment: Towards a Policy of Strong Sustainable Development". In Biscontin, G., Driussi, G. (Eds.), Proceedings of Scienza e Beni Culturali, Vol. 30, Quale Sostenibilita' per il restauro?, Brixen, 1–4 July 2014, Venice: Edizioni Arcadia Ricerche: 457–467.

Vandesande A., Michiels E. & Van Balen K. (2016). "From a working definition of preventive and planned conservation towards the innovative services of Monumentenwacht Vlaanderen". In Della Torre, S. (Ed.), Borgarino, M. (Ed.), *Proceedings of the International Conference Preventive and Planned Conservation: Vol. 4*. International Conference Preventive and Planned Conservation. Monza-Mantua, 5–9 May 2014 (pp. 25–34). Milan: Politecnico di Milano e Nardini Editore.

Vandesande, A. & Van Balen, K. (2016). "An operational preventive conservation system based on the Monumentenwacht model". In: Van Balen, K., Verstrynge, E. (Eds.), Proceedings of the 10th International Conference on Structural Analysis of Historical Constructions (SAHC2016), Anamnesis, diagnosis, therapy, controls. Leuven (Belgium), 13–15 September 2016, Leiden: CRC/Balkema: 217–224.

Waller, R. (2003). 'Cultural property risk analysis model', Göteborg studies in conservation, 13: Göteborg: Acta Universitatis Gothoburgensis.

Wordsworth, P. (2001). Lee's Building Maintenance Management, London: Blackwell Science.

Worthing, D. & Bond, S. (2008). Managing Built Heritage: The Role of Cultural Significance. Oxford: Blackwell.

Yau, Y. (2012). 'Multicriteria Decision Making for Homeowners' Participation in Building Maintenance', Journal of Urban Planning and Development, 183 (2):110–120.

Zavadskas, E., Kaklauskas, A. & Banaitis, A. (2010). 'Real estate's knowledge and device-based decision support system', International Journal of Strategic Property Management, 14 (3): 271–82.

Innovative Built Heritage Models – Van Balen & Vandesande (Eds)
© 2018 Taylor & Francis Group, London, ISBN 978-1-138-49861-7

Risk management and built heritage: Experiences, reflections and perspectives

A. Konsta
Politecnico di Milano, Milan, Italy

ABSTRACT: In recent years the importance of risk assessment within the practice of condition surveys in the field of built heritage conservation has been gaining ground, thus acknowledging the positive results of a risk management approach, consolidated as a practice, in the field of museums and their collections. This study provides a review of recent experience, such as recommendations and guidelines, best practices and projects, that address risk management methodology for the preventive and planned conservation of built heritage, both in the Italian and European context. Its purpose is to understand the different approaches and to analyse the methods and models defined by these projects, in order to form an operational reference framework which, in its turn, will be able to contribute to the development of a holistic risk management approach to built heritage.

1 INTRODUCTION

In Ortwin Renn's words (1992) risk is defined as the possible harm caused by natural phenomena or human activities, which may happen. It is based on a twofold concept: one descriptive and the other normative. The term risk, on the one hand, analyses the cause-effect relation and, on the other hand, implies the possibility of the avoidance or mitigation of the effects by reducing the causes or the consequences respectively (Renn 1992).

This kind of reasoning is at the base of conservation practice and its effort to understand deterioration processes so as to bring them under control (Grandori 1985, Della Torre 2010).

The first part of this study concerns the gradual steps made and the results achieved by different experiences in accordance with this line of research, during the last few decades, both in the Italian and European context. The intention is to put them into groups according to their common aspects and then, through an analysis of the most relevant cases, to reflect upon the critical issues and to reach conclusions about the relevance of a risk management approach to built heritage (Table 1).

2 REVIEW OF RECENT EXPERIENCES

2.1 *Risk management and preventive conservation*

From the 1970s onwards the importance of a scientific approach to conservation issues has been recognised, making it necessary to ground conservation techniques objectively. To this end, research in the field of conservation has focussed on the study of the nature of materials, the causes and extent of their decay, as well as their environmental conditions, and has subsequently developed into a series of technical regulations, from the 1980s at a national level (Commissione NorMal and Commissione UNI Beni Culturali—NorMal) and from the dawn of the new millennium at an international European level (European Committee for Standardization CEN-TC 346 Conservation for Cultural Property), where the contribution of the Italian experience is of particular influence (Alessandrini 2011).

Specifically, the European activity of standardization in the field of conservation regards terminology and definitions concerning movable and immovable cultural property, methods for an examination of materials and their deterioration processes, with regard to the environmental conditions, as well as products and technologies used in conservation work (Fassina 2010).

The scientific knowledge acquired has, on the one hand, led to more appropriate methods of intervention and curative approaches and, on the other hand, given rise to regular inspection surveys, to systematic monitoring, to periodic maintenance and risk assessment practices as the most proper preventive strategies. Indeed, the European standard *Conservation of cultural property—Condition survey of immovable heritage* (EN 16096:2012) despite presenting some methodological problems, represents an attempt to implement preventive strategies (Della Torre 2010). Another significant contribution in this direction is the European Cultural Heritage Identity

Card—EU-CHIC project, a proposal for a protocol for integrated documentation of tangible cultural heritage (2009–2012), which provides methods and tools for systematic collection and storage of data and for tracking environmental changes and human interventions, as well as developing criteria and indicators for risk assessment.

Furthermore, technological progress has contributed to the development of more accurate diagnosis methods, of innovative monitoring systems and of advanced damage assessments. Some representative examples are the recent European projects: SMooHS (2008–2012) dealing with the direct tracing of damage, TEACH (2008–2012) with regard to the impact assessment on historic buildings caused by the main pollutants, FIRE-SENSE (2009–2012) relative to early warning monitoring in case of fires and storms, NOAH'S ARK (2006–2007) and PARNASSUS (2010–2013) concerning the climate change impacts on cultural heritage, and PERPETUATE (2010–2012) apropos of structural monitoring and seismic vulnerability assessment based on advanced performance based models.

All of the above demonstrate the efforts made to measure environmental parameters, to understand and assess their impacts on historic buildings and to evaluate risks in order to prevent, control and mitigate damages.

In addition, the introduction of the carrying capacity and equivalent safety (Italian safety legislation, 1998) concepts, based on a rational risk management approach (Della Torre 2001), as well as the seismic improvement notion (PCM Recommendations, 2011) represent considerable shifts towards a preventive conservation practice rooted in a thorough knowledge of the building and its environmental conditions.

2.2 *Risk management and planned conservation*

Moreover, during the 1970s the signs of inadequate urban planning and countryside abandonment that had been manifested on a territorial scale in the degradation of built heritage and the rural landscape, and in the acceleration of natural disasters, had made it clear that the problem of architectural conservation could not be addressed separately from its physical environment (Urbani 2000, Zanardi 1999) These concerns prompted Urbani's efforts towards repositioning conservation matters within the sphere of cultural property (Petraroia 2000, Della Torre 2010). For this reason his research, carried out in the Umbria region, but with an eye to the feasibility of implementing it throughout the whole of Italy, identified technical tools for drawing up an intervention plan based on a quantitative analysis through the evaluation of the risk factors and their consequences, related to materials and structural typologies (Urbani 2000).

Later, and pursuing this line of research, came the Risk Map project, coordinated by the Central Restoration Institution in 1992, followed by the Guidelines for Conservation Plan, approved by the Lombardy Region and published in 2003. Both these practical tools seek to promote the culture of prevention and maintenance and to implement a risk management strategy for the built environment on a territorial and building scale.

In a similar way, in the international context, the increasing pressure to respond to emergencies, provoked either by natural disasters or human destruction, has prompted the conservation movement to focus on prevention and adopt a "*cultural-heritage-at-risk*" framework (Stovel 1998). The most important international organizations, such as UNESCO, ICOMOS, ICCROM, Council of Europe, and many others, have strengthened their activities to develop operative methods for an overall prevention and mitigation strategy for the protection of cultural heritage (Feilden 1987, Stovel 1998). The *Risk-Preparedness Manual for World Cultural Heritage* in 1998 and *Between Two Earthquakes* in 1987 are two of the first publications in a series of handbooks that is still being added to and updated in view of current concerns related mainly to the risks of climate change faced by cultural heritage (e.g. *Climate Change and World Heritage Report on predicting and managing the impacts of climate change on World Heritage and Strategy to assist States Parties to implement appropriate management response*, 2007).

2.3 *Value-based risk management and sustainable conservation*

From the 1990s onwards the idea began to crystallise that sustainable development is strongly related to the conservation of cultural heritage. In the light of this, conservation issues now stretch to embrace social, cultural and economic development objectives (Della Torre 2006).

Thus Della Torre highlights the importance of recognising the co-evolution potential of cultural heritage, drawing attention to its vital (social and economic) context and the interactions that can be created in the course of time. In other words, the aim of conservation is not to halt the evolution process by isolating the "object" from its reality, but instead to manage its transformation by means of systemic projects (e.g. cultural districts) and preventive strategies, varying from the environmental scale to use practises and to valorisation aspects (Della Torre 2006).

An analogous approach was taken by the programme Heritage at Risk initiated by English

Heritage in 1984, for "*typical historic buildings types suffering redundancy and neglect as symptoms of the economic rationalisation of the birthplace of the industrial revolution*". Finding profitable alternative uses or new users has offered the best opportunities for positive development ensuring the repairs and survival of heritage assets (Filder 1987).

Another significant example is the Halland Model, developed in Sweden during the recession of the 1990s. Initially, the three mottos of the Halland model were "save the jobs" (235 jobs created), "save the craftsmanship" (1200 construction workers) and "save the buildings" (100 historic buildings at risk). A later addition was "new uses for the improved premises" (Gustaffson 2009). As Gustaffson points out the challenge was to formulate a new strategy, based on a multi problem-orientated approach, for sustainable development, by being aware of the potential role of cultural heritage.

Moreover, during the mid-1990s and 2000s, conservation practice in the United Kingdom introduced value-based heritage management built on the Burra Charter process and the significance of place to inform decision-making (Clark, 2014). The Conservation Management Plans developed by English Heritage and the Heritage Lottery Fund, the English Heritage publication *Power of Place* (2000), as well as its *Conservation Principles, Policies and Guidance for the Sustainable Management of the Historic environment* (2008), and the National Trust "Going Local" strategy (2010) are some of the most important projects to incorporate values in conservation (Clark 2014). The principal elements of these plans number three: first, the assessment and articulation of cultural significance; second, the vulnerability assessment of significance and, third, the development of policies and practices for the mitigation of vulnerability (Worthing 2008, Dann 2014).

At this point it is important to mention the national plan for risk prevention launched in 2016 in Italy, called "Casa Italia". This large-scale project aims, on the one hand, at defining a general risk preparedness strategy able to ensure the safety of people and of their built environment and, on the other hand, at evaluating the opportunities presented. The integrated approach combining diverse risks as well as opportunities is based on a systemic gathering of available data and the implementation of best practices, matured in recent years (PCM, 2017).

To summarise, during the last four decades risk management methodology, based on identification, analysis, assessment and mitigation of risks, has been recognised as a more rational and transparent approach to the conservation decision-making process.

Organising them into three groups does not imply a separation of different schools of thought but traces an evolution in regard to three major areas of concern, firstly the importance of scientific knowledge towards a better understanding of material deterioration and what causes it; secondly the need to transcend the fragmentary vision of "monuments" for a broader and more integrated study of cultural heritage and its conservation; and thirdly the urgency to find a more sustainable way to achieve this.

3 RISK MANAGEMENT OBJECTIVES AND APPROACHES

There are three main elements that distinguish the objectives and the approaches of the different experiences in risk management indicated above: the first is the number (one risk, combination of risks, all possible risks) and the type of risks examined and assessed (slow/cumulative risks or sudden/catastrophic ones). The second is the "object" that is taken into consideration each time, or else the scale dimension of the analysis, a building scale or a territorial scale. The third is its "context", and to what extent its various physical, cultural, social, and economic aspects are integrated.

Many experiences of dealing with the assessment and the mitigation of seismic risk (Italian guidelines, PERPETUATE project, etc.) can be included in the category of the risk-based approach which takes only one risk into account, both at a territorial and at a building level. During the 1990s the devastating impacts of earthquakes on human lives and the extensive damage to historic buildings led, especially in countries in seismic zones, to ongoing research activity aimed at an analysis of the seismic risk to cultural heritage through research on masonry behaviour and procedures to assess vulnerability.

Instead, the Risk Map project, as well as the recent national Casa Italia plan, are examples of risk assessment approaches that cope with two or more different risks. Whereas the former examines the principal risk factors, classified into three main categories (geomorphologic, climatic/pollution, socio-economic development) connected respectively to the physical, structural and non structural aspects, and occupancy/safety aspects of the building, the latter takes into account the combination of the two main seismic and hydrogeologic risks that repeatedly threaten a large part of the country, associated with the physical aspects of the buildings and the socio-economic aspects of the population.

Risk typology in some sense determines the dimension scale of the "object" taken into account.

Sudden/catastrophic risks, such as landslides, river floods, earthquakes, etc., are mainly related to the territorial scale and their mitigation, if possible, requires interventions of this dimension. Moreover, the fact that they have an overall impact make them more suitable to be dealt with on this scale, making planning more efficient by identifying priorities and ensuring a better use of resources. Risk analysis carried out at this level is based on the collection of data concerning either the hazards or their consequences (hazard maps, vulnerability maps, risk maps). Whereas the detection and estimation of many slow and cumulative risks, of a natural or human nature, such as biological, meteorological, or of poor maintenance, inadequate use, inappropriate standards of comfort, and many others, are much dependent on a building scale survey. The "Monumentenwacht", as well as the "Maintain our Heritage" approach in the Northern Europe are indicative cases of this category.

As for the extent of the integration of different perspectives regarding "context" we can notice the gradual growth of interdisciplinary approaches. If, during the 1980s and 1990s, risk-based analysis was mainly characterised by a technical approach based on quantitative and objective data, determined by scientific measurements, from 2000 on it is more open to social-cultural and economic considerations. Research activity connected with materials and their deterioration processes in regard to environmental conditions, as well as an analysis of constructive techniques and an assessment of their vulnerability, has a technical approach that only takes into account the physical aspects. On the other hand, the line of research related to systemic relations between cultural heritage and its context has drawn attention to society, by reflecting on the role and participation of the users/stakeholders. For example, Monumentenwacht Vlaanderen has involved users in daily maintenance, Lombardy Cultural Districts has supported capacity building in its territory, and the Halland Model has encouraged regional cooperation between the labour market, the construction industry and the cultural heritage sector.

4 RISK COMPONENTS AND ASSESSMENT METHODS

As Renn (1992) underlines, despite the variety of risk concepts described in the literature of various (technical, economic, social, etc) disciplines, there is a common element that unifies the various approaches and definitions and that is the distinction between reality and possibility. Thus, the term risk indicates "*the possibility that an undesirable state of reality (adverse effects) may occur as a result of natural events or human activities*". From this perspective, the connection between causes and effects cannot be predetermined; that is to say, the concept of risk is not based on the principle of linear causality, where the effect is determined by the cause (same cause—same effect), but instead suggests a consideration of the particular circumstances presented each time (same cause—different conditions—different effect).

This idea is central to the widely accepted definition of risk as a function of two principal, dominant components: "hazard" and "vulnerability". The former term regards the characteristics of a particular phenomenon that could cause damage, while the latter is related to the characteristics of the exposed system and its propensity to damage. To assess risk, it is necessary to evaluate these two parameters which, in their turn, are functions of other variables.

Concerning hazards, built heritage is exposed to both natural and human risk factors. The former category comprises events such as earthquakes, landslides and avalanches, floods, volcanic eruptions, etc; meteorological factors, such as solar radiation, air humidity, wind, tornadoes, hurricanes; biological factors, like lichens, mosses, animals; and geochemical factors, such as soluble salts and ground erosion. The latter category, related to human activities, includes pollution, inappropriate use and interventions, inadequate management, as well as vandalism, theft, war, and other economic and social factors that can have strong effects on cultural heritage (Camuffo 1997).

To assess hazards it is generally necessary to estimate three principal aspects: probability/frequency of occurrence, intensity, and spatial extension of the phenomenon. According to the specific characteristics of hazards the assessment process may include scientific studies, historic analysis, modelling, experience-based knowledge, experimentation and testing (Poljanšek et al 2017). Although we cannot pretend to understand all about hazards and their potential impacts (Poljanšek et al 2017), continuous scientific research and advances in technology are teaching us more and more about them. In particular, current studies in this field concern the interactions between hazards and the chain of events that is able to propagate the consequences and increase the total risk.

As to vulnerability assessment, it is less clear and not so evident which indicators have to be selected and measured and hence we find a variety of interpretations and, consequently, methods of evaluation in the relative literature. For example, the NOAH'S ARK project estimates vulnerability according to the intrinsic characteristics of building materials and their position (indoor, outdoor, buried) to climate variables; Risk Map project assesses the state of conservation of surfaces,

Table 1. Reference framework of the different approaches to risks and their assessment methods related to the built cultural heritage.

Case studies	Approach to risks	Approach to vulnerability	Vulnerability indicators	Risk assessment methods
PCM Recommendations for the assessment and mitigation of seismic risk to cultural heritage	single-risk approach: earthquakes	territorial scale: physical	material, dimensional and structural characteristics, location, morphological aspects, occupancy and eventual overcrowding	(LV1) seismic safety evaluation: qualitative approach based on survey method and simplified evaluation models.
		building scale: physical		(LV2) evaluation of individual macro elements: local collapse mechanisms, (LV3) overall evaluation of the building seismic response: global model.
Risk Map project	multi-risk approach: (H1) geophysical, hydrogeological hazards, (H2) climatic, air pollution and (H3) socio-economic risk factors.	territorial scale: systemic	(VI) constructional and static-structural characteristics, (V2) surface aspect, (V3) use and safety aspects.	risk is expressed as a general function of the vulnerability components, related to each architectural or archaeological immovable item and of danger, related to each territorial area on which the item is located.
PARNASSUS project	multi-risk approach: combination of two risks related to climate change: floods, wind driven rain.	building scale: systemic	age, listed status, use, footprint and number of storeys, material and structural characteristics, condition.	risk assessment based on damage functions: direct tangible (physical damage) and direct intangible damage are considered.
Noah's Ark project	multi-risk approach: combination of risks related to climate change and pollution.	territorial scale: physical	material type and location.	risk and multiple risk maps based on material damage functions and dose-response functions related to the changes of the most critical meteorological and pollution parameters to the built cultural beritage.
Climate for Culture project	multi-risk approach: combination of risks related to climate change.	building scale: physical	material type and building type (building volume, window area, structure weight, moisture buffering performance).	outdoor climate risk maps and indoor climate risk maps based on damage functions, including the synergistic effect of various climate and chemical variables, thresholds and cycles.

(Continued)

Table 1. (*Continued*).

Case studies	Approach to risks	Approach to vulnerability	Vulnerability indicators	Risk assessment methods
SMooHS project	multi-risk approach: risks of long term influence.	building scale: physical	material and structural characteristics, condition, use.	direct tracing damage models. material and structural deterioration models that into account the data from continuous monitoring.
Monumentenwacht Vlaanderen (interior assessment), Museum Amstelkring, Petra archaeological park	multi-risk approach: identification of all possible risks.	building/ site scale; systemic	those aspects (material, structural, condition, use, etc.) that determine the vulnerability of the different elements to the specific risks identified.	risk assessment method based on the model defined by R. Waller: the magnitude of risk is the product of the probability of occurrence (or extent) and severity of effect: Severity is the product of the fraction of the collection susceptible to the threat and the potential loss of value to the collection.
Cultural Districts, Halland Model, Heritage at risk	multi-risk approach: social and economic risk factors.	territorial scale: systemic	occupancy and condition of buildings, cultural, social and economic indicators.	qualitative approach based on survey method and context analysis.

structures and the occupancy/safety aspects of the building; Heritage at Risk evaluates the occupancy and the physical conditions of the buildings; the "equivalent safety" concept is based on an evaluation of the compatibility of the ways the building is used; and Casa Italia considers the social and economic conditions that are present.

In order to quantify risk, in terms of expected damages, it is necessary to calculate the dimension of the exposed system (number, area, or other units of measure). The term used commonly in literature to indicate this parameter is "exposure", although in some cases it is seen to be integrated within an assessment of the vulnerability component (e.g. MOVE—Methods for the Improvement of Vulnerability Assessment in Europe). However, the most standard form of risk assessment is the one that includes all three components separately: R = f (Hazard, Exposure, Vulnerability).

What becomes very relevant to the case of cultural heritage is how to evaluate damages and choose the most appropriate criteria for identifying priorities. Indeed, the attribution of values to the physical properties of heritage has always been at the base of heritage conservation. Over the last few years a recognition of the multivalent character of heritage and the mutable and complex nature of its values (ICOMOS Burra Charter, 1979; Nara Document on Authenticity, 1994; Getty Conservation Institute, Research report Assessing the values of Cultural Heritage, 2002) has raised the question of an interdisciplinary approach and led to more systematic methods for their identification and assessment (De la Torre & Mason 2002, Van Balen 2008). Despite the methodological difficulties presented by a value assessment procedure, caused by the diverse and sometimes conflicting nature of values as well as by their mutability, consequent on their context dependence (Mason 2002), incorporating it into a risk assessment process becomes indispensable for an evaluation of impacts from different perspectives and an understanding of the loss of value.

5 CLASSIFICATION OF RISKS

The methodology of risk assessment described above has mainly been implemented on a territorial scale and has principally been connected with catastrophic risks. On the scale of architecture, risk analysis has been incorporated more implicitly, rather than explicitly, into an analysis of causes of decay, thereby restricting the process of developing a methodology of risk assessment in the field of built heritage (Della Torre 2010).

In contrast, in the field of museums risk assessment practices have been implemented from the 1990s onwards and have developed their own methodology, based on a system for grouping risks in categories (Waller 1994). This classification took two elements into consideration, firstly the agent of deterioration related to the risk (ten agents of deterioration have been identified by Michalski, nine relative to physical deterioration and one concerning non-physical deterioration) and, secondly, the type of risk, according to the frequency of occurrence and the severity of effect on collections (three types of risks were recognised, rare/catastrophic, sporadic/severe, constant/mild, gradual). Each category of risk includes risks of the same type connected with the same agent of deterioration. What becomes relevant, as Waller (1995) explains, is that a knowledge of types of risk can help to clarify different strategies for their assessment. Where information regarding risk magnitude is inaccurate or lacking research plans can be developed in order to address priority issues. Even if it proves impossible to accurately estimate all risks, Waller (1995) underlines the importance of a holistic approach, which takes into account not only the most severe risks but also the less dangerous, for an understanding of the risk variety and hence more rational decision making.

Only in recent years has it been attempted to apply and adapt this methodology to the field of built heritage. Specifically, to historic buildings and their collections (e.g. risk assessment of Museum Amstelkring, 2003), to the interiors of historic buildings (e.g. Monumentenwacht Vlaanderen, 2007), and to archaeological sites (e.g. Petra archaeological park, 2011). All three of them have implemented the Cultural Property Risk Analysis Model developed by Waller and have offered new perspectives producing significant results that have shaped a more systematic approach to built heritage risk management.

Although the limits presented in this "agent-based" approach of risk assessment (Taylor in his article *Distinguishing between the map and the territory* points out the difficulty in representing the synergistic action of the different agents) the importance of a risk assessment methodology is prominent in the systematic analysis and division of hazards into categories and the consequent articulation of different levels of control and strategies for their mitigation.

This is particularly significant for the scale of architecture which is interrelated to the scale of urban and territorial planning. Many examples that have been taken into consideration in the first part of this study (Pilot project of Umbria, Cultural Districts, Casa Italia, etc.) on the one hand, demonstrate the development of this awareness and, on the other hand, indicate the encouraging results of the implementation of this type of evaluation.

6 CONCLUSIONS

In short, the complexity that characterises the scale of architecture and consequently the risks related to it – variety of time and spatial scale, of context dimensions – has led to the development of different forms of risk analysis and evaluation related to specific risks, more or less formalised according to the severity of their impacts, and has resulted in the sharing of objectives and the finding of "common ground".

On the other hand, the awareness of this complexity has imposed the need to interrelate the different scales. This has led to positive and concrete results that can contribute decisively to a holistic risk management approach to built heritage. In this perspective, the importance of the complex interactions between cultural heritage and its social and economic context cannot be ignored, giving rise to the consideration of a systemic approach to vulnerability and the development of strategies for its reduction.

But, what is the most appropriate approach to the integration of different level strategies (building and territorial scale)? How the analysis of risks at these two different scales can contribute to a holistic approach to the management of built heritage? Taking into consideration, on the one side, the relevance of specific aspects which characterise a building scale to an intervention/conservation planning, and on the other side, the necessity of an analysis at territorial level for an understanding of the interdependencies between building and its context, it becomes clear the importance of the integration of these two analyses in a more standardised procedure.

Indeed, the classification of risks and the development of a systematic method of analysis, assessment and management of risks offer a way to integrate multilevel strategies, interdisciplinary collaboration, and an exchange of knowledge, indispensable for a coherent preventive conservation practice. This procedure is essential for the creation of an operational framework, able to guarantee a rational process of decision making and an efficient allocation of the resources for the management of the built heritage.

REFERENCES

Ashley Smith, J. (2013). *Report on damage functions in relation to climate change*, Climate for Culture.

Alessandrini, G. 2011. La conservazione dei beni culturali e la normativa tecnica: passato, presente e futuro. *U&C*, 7: 27–42.

Bertolin, C. & Camuffo, C. (2014). *Report on Climate change impact on movable and immovable cultural heritage throughout Europe*, Climate for Culture.

Brokerhof, A., Luger, T., Ankersmit, B., Bergevoet, F., Schillemans, R., Schoutens, P., Miller, T., Kiers, J., Meuthing, G. & Waller, R. (2005). Risk Assessment of Museum Amstelkring: Application to an historic building and its collections and the consequences for preservation management. In: *Preprints of the 14th ICOM-CC Triennial Meeting, The Hague 12–16 September 2005*: 590–596. London: International Council of Museums and James & James Publishers.

Camuffo, D. (1997). Perspectives on risks to architectural heritage. In: N.S. Baer & R. Snethlage (eds), *Saving our architectural heritage: the conservation of historic stone structures*: 63–92. New York: John Willey & Sons Ltd.

CEN/TC 346 – Conservation of Cultural Heritage, EN 16095:2012 (WI = 00346012), *Conservation of cultural property—Condition recording for movable cultural heritage.*

Clark, K. (2014). Values-Based Heritage Management and the Heritage Lottery Fund in the UK. *APT Bulletin: The Journal of Preservation Technology, Special issue on values-based preservation*, 45 (2/3): 65–71.

Dann, N. (2014). Planned and preventative conservation and England: experiences, opportunities and requirements. In: *Preventive and planned conservation: proceedings of the international conference. Vol 4: Metodi e strumenti per la prevenzione e la manutenzione, Monza—Mantova, 5–9 May 2014*. Firenze: Nardini, pp. 1–10.

De la Torre, M. (2002). *Assessing the values of cultural heritage: Research Report*. [online] CA: Getty Conservation Institute: 3–4. Available at: http://hdl.handle.net/10020/gci_pubs/values_cultural_heritage [Accessed 31 May 2017].

Della Torre, S. (2001). Dal bisogno della sicurezza alla cultura della prevenzione. *Tema: Tempo, Materia, Architettura*, (1): 3–5.

Della Torre, S. (2006). Programmare la conservazione: valore culturale e sostenibilità. In: *La fruizione sostenibile del bene culturale: atti del convegno, Firenze 17 giugno 2005*: 24–27. Firenze: Nardini.

Della Torre, S. (2010). Preventiva, integrata, programmata: le logiche coevolutive della conservazione. In: *Pensare la prevenzione. Manufatti, usi, ambienti: atti del XXVI convegno Scienza e Beni culturali, Bressanone 13–16 luglio 2010*: 67–76. Venezia: Arcadia Ricerche.

Della Torre, S. (2010). *Critical reflection document on the draft European Standard CEN/TC 346 WI 346013 Conservation of cultural property—Condition survey of immovable heritage* [unpublished discussion document].

Della Torre, S. (2014). Oltre il restauro, oltre la manutenzione. In: *Preventive and planned conservation: proceedings of the international conference, Monza, Mantova, 5–9 May 2014*:1–10. Firenze: Nardini.

Fassina, V. (2010). European Technical Committee 346: Conservation of cultural property: Updating of the activity after a three year period. In: *Wood science for conservation of cultural: proceedings of the international conference held by COST action IE060, in Braga 5–7 November 2008*: 251–258. Firenze: Firenze University Press.

Feilden, Bernard M. (1987). *Between two earthquakes: cultural property in seismic zones*. Rome; Marina del Rey, CA: ICCROM; Getty Conservation Institute.

Fidler, J. (1987). Historic Buildings at Risk: a strategy to identify redundant (under-used) and dilapidated buildings and to rescue them. In: *Old cultures in new worlds. 8th ICOMOS General Assembly and International Symposium*: 884–891. Washington: US/ICOMOS.

Guagenti Grandori, E. (1985). Strategie di controllo nella tutela dell'edificato. In: A. Bellini (ed), *Tecniche della conservazione*: 79–96. Milano: Franco Angeli.

Gustafsson, C. (2009). Modelling experiences from regional development and learning districts using built cultural heritage and collaborative management research. In: F. Putignano (ed), *Learning Districts, Patrimonio culturale, conoscenza e sviluppo locale*: 79–102. Santarcangelo di Romagna: Maggioli editore.

International Council on Monument and Sites (ICOMOS), 2013. *The Australia ICOMOS Charter for the Places of Cultural Significance*.

Lemaire, R. & Stovel, H. (eds.) (1994). *The Nara Document on Authenticity*, Nara Conference on Authenticity in Relation to the World Heritage Convention. Nara, Japan, 1–6 November 1994.

Mason, R. (2002). Assessing values in conservation planning: methodological issues and choices. In: M. De la Torre (ed), *Assessing the values of cultural heritage: Research Report*: 5–30. [online] CA: Getty Conservation Institute.

Available at: http://hdl.handle.net/10020/gci_pubs/values_cultural_heritage [Accessed 31 May 2017].

Meul, V. & Stulens, A. (2010). The integral approach of Monumentenwacht Vlaanderen: a model for implementing participatory preventive conservation for historic interiors. In: *Multidisciplinary conservation: a holistic view for historic interiors: proceedings of the joint interim meeting, Rome 23–26 March2010*: 1–11. Rome: ICOM-CC.

Michalski, S. (1990). An overall framework for preventive conservation and remedial conservation. In: *ICOM Committee for Conservation, 9th triennial meeting, Dresden, German Democratic Republic, 26–31 August 1990, preprints:* 589–591. ICOM Committee for Conservation.

Paolini, A., Vafadari, A., Cesaro, G., Santana Quintero, M., Van Balen, K., Vileikis, O. & Fakhoury, L. (2012). *Risk Management at Heritage Sites: a case study of the Petra World Heritage Site*, Paris: UNESCO and Katholieke Universiteit Leuven, Faculty of Engineering, Raymond Lemaire International Centre for Conservation.

Petraroia, P. (2000). Postfazione. In: B. Zanardi (ed), *Intorno al restauro*: 153–157. Milano: Skira.

Poljanšek, K., Marin Ferrer, M., De Groeve, T., Clark, I., (eds.) (2017). *Science for disaster risk management 2017: knowing better and losing less. EUR 28034 EN*. Luxembourg: Publications Office of the European Union.

Presidenza del Consiglio dei Ministri (PCM), Struttura di Missione Casa Italia, 2017. *Rapporto sulla Promozione della sicurezza dai Rischi naturali del Patrimonio abitativo*.

Presidenza del Consiglio dei Ministri (PCM). (2011). *Recommendations for the assessment and mitigation of seismic risk to cultural heritage with reference to*

the Italian Building Code, NTC2008. Directive of the Prime Minister, 9/02/2011. G.U. no. 47, 26/02/2011, suppl. ord. no. 54.
Renn, O. (1992). Concepts of risk: a classification. In: S. Krimsky (ed), Social theories of risk: 53–79. Westport, Conn.: Praeger.
Stephenson, V. & D'Ayala, D. (2014). A new approach to flood vulnerability assessment for historic buildings in England, *Nat. Hazards Earth Syst. Sci*, 14: 1035–1048. Available at: www.nat-hazards-earth-syst-sci.net/14/1035/2014/.
Stephenson, V., D'Ayala, D., & Aktas, Y.D. (2016). Assessment of flood and wind driven rain impact on mechanical properties of historic brick masonry, in *International RILEM Conference on Materials, Systems and Structures in Civil Engineering Conference segment on Historical Masonry 22–24 August 2016, Technical University of Denmark, Lyngby, Denmark*.
Available at: https://www.researchgate.net/publication/306379378.
Stovel, H. (1998). *Risk preparedness: a management manual for world cultural heritage*. Rome: ICCROM.
Taylor, J. 2012. Distinguishing between the Map and the Territory: Synergy in Agent based Approaches to Risk Assessment, R. Waller (ed), The International Symposium on Cultural Property Risk Analysis, a focus issue of *Collections: A Journal for Museum and Archives Professionals*, 8(4): 297–306.
Urbani, G. (2000). In: B. Zanardi (ed), *Intorno al restauro*: 103–112. Milano: Skira.
Van Balen, K. (2008). The Nara grid: an evaluation scheme based on the Nara document on authenticity. *Association for Preservation Technology International (APT Bulletin) Special Issue*, 39 (2/3): 39–45.
Waller, R. (1994). Conservation risk assessment: a strategy for managing resources for preventive conservation. *Studies in Conservation*, 39 (sup2): 12–16.
Waller, R. (1995). Risk management applied to preventive conservation. In: C.L. Rose, C.A. Hawks, H.H. Genoways (eds), *Storage of natural history collections: A preventive conservation approach*: 21–27. New York: Society for the preservation.
Worthing, D., Broström, T., Jörgensen, M.H., Hansson, J., Tooming K. & Vainlo, A. (2013). *Guidelines for sustainable management of churches Sustainable Management of Churches—Conservation Management Plan Guideline: Research report*. [online] Estonia: Conservation Centre Kanut.
Available at: http://smcproject.org.ee/__smc_raports/WP2/guidelines/SMC_WP2_Guidlines_online.pdf.
Zanardi, B. (1999). Le attività di tutela dopo il 1963. In: B. Zanardi (ed), *Conservazione, restauro e tutela, 24 dialoghi*: 9–52. Milano: Skira.

Websites

cartadelrischio.it. Risk Map official website. [on line] Available at: http://www.cartadelrischio.it/eng/dati_vuln.html [Accessed 10 Oct. 2016].
cordis.europa.eu. Firesense project [on line] Available at:http://cordis.europa.eu/project/rcn/93948_en.html [Accessed 15 Oct. 2016].
cordis.europa.eu. Teach report summary [on line] Available at: http://cordis.europa.eu/result/rcn/162482_en.html [Accessed 10 Nov. 2016].
eu-chic.eu. Eu-chic project official website. [on line] Available at: http://www.eu-chic.eu/ [Accessed 10 Oct. 2016].
ipcc.ch. Climate change report. [on line] Available at: http://www.ipcc.ch/report/ar5/ [Accessed 15 Nov. 2016].
move-fp7.eu. MOVE project official website. [on line] Available at: http://www.move-fp7.eu [Accessed 20 Oct. 2016].
perpetuate.eu. Perpetuate project official website. [on line] Available at: http://www.perpetuate.eu/ [Accessed 15 Oct. 2016].
ucl.ac.uk. Noah's Ark project [on line] Available at: https://www.ucl.ac.uk/bartlett/heritage/research/projects/project-archive/noahs-ark-project [Accessed 10 Nov. 2016].
ucl.ac.uk. Parnassus project [on line] Available at: https://www.ucl.ac.uk/parnassus [Accessed 10 Oct. 2016].
whc.unesco.org. Climate change. [on line] Available at: http://whc.unesco.org/en/climatechange/ [Accessed 15 Nov. 2016].

Innovative Built Heritage Models – Van Balen & Vandesande (Eds)
© 2018 Taylor & Francis Group, London, ISBN 978-1-138-49861-7

Preventive actions for architectural heritage against seismic risk. The Italian experience

L. Cantini
Politecnico di Milano, Milan, Italy

ABSTRACT: The difficult management of historical buildings subjected to seismic events represents an open question, as shown by the recent sequence of earthquakes stroking the center of Italy. The fundamental assumptions of the conservation design, based on the protection of the tangible and intangible values of the historical buildings, meet a difficult field of application when the material components of the architectural heritage are seriously compromised. In this case, the strategies for the prevention from seismic risk is the correct solution for a serious protection policy addressed to the built heritage. This work focuses on the limits and on the positive results obtained by preventive actions carried out in past episodes on the Italian territory, considering the roles set by several experts, in past and recent times, for facing the problem of seismic risk connected to cultural heritage safeguard.

1 INTRODUCTION

The conservation design for historical buildings placed in seismic areas is a borderline topic, traditionally field of interest shared by several competencies. Architects and engineers provided several theoretical supports for studying historical buildings from the point of view of the vulnerability of their structures and therefore the identification of the preventive interventions requested for increasing the safety of the building system. In the wide experience of the Italian seismic events, they assumed a role of connection between the conservation demands, coming from historians, and the improvement of the building global response requested by other experts, like geotechnics and civil engineers. This long tradition drove to a peculiar organization devoted to the protection of the cultural heritage set in seismic areas, characterized by a deep collaboration between experts from different fields. As a result, a special national apparatus was set for supporting the studies carried out by those experts in terms of risk and post-earthquake management. Institutions like the National Civil Protection or the Large Risks Commission were introduced in the Italian protection system in order to provide a rapid response after catastrophic events (like earthquakes) and to collect information for identifying resources and vulnerabilities of historical buildings.

In few years, since the L'Aquila earthquake in 2009 to the recent seismic sequence in Amatrice and Norcia in 2016, several historical centres in Italy were stroke by important natural events. Each time, after the distressing damage quantification, in terms of killed people and lost built heritage, the debate on the vulnerability of the historical centres becomes prevailing. Also after the recent devastating earthquake that produced deep damages in Amatrice and Norcia areas, the idea of preventive actions on the architectural heritage for the seismic risk reduction is assumed as a topic in public discussion and commented by various experts. This work provides an overview of the evolution of the preventive measures, at the level of codes and guidelines, set for protecting the architectural heritage, defined as:

- identification of the historical buildings characteristics;
- evaluation of their structural systems;
- recognition of local strategic solutions for mitigating the seismic effects;
- estimation of the buildings vulnerability.

This wealth of knowledge should drive the choices for the interventions following clear theoretical assumptions. However, the difficulties for applying a shared methodology, left space to unreliable decisions, where the expertise matured in the long tradition of anti-seismic preventive actions is not recognizable. Rather than promoting a less invasive improvement, limited to vulnerable structural elements, the idea of an invasive retrofitting of the building structures remains the most reliable answer for many professionals involved in the problem.

2 PREVENTIVE MEASURES FOR SEISMIC RISK MITIGATION DEVELOPED IN THE PAST

2.1 *Historical measures against seismic vulnerability*

The Italian territory has been stroke by severe earthquakes in the past. Studying the problem from an historical perspective, different preventive measures were introduced after various seismic events. In some cases, these anti-seismic solutions were codified in specific treatise. Since the appearance of the building techniques codified by the Romans, to the detailed recommendations for the realization of antiseismic structures provided in specific acts of the Kingdom of Naples (including from 13th to 18th century the south part of the peninsula), the problem of the vulnerability of the buildings to seismic events has been taken into account. Appealing solutions like masonry walls integrated by wooden frames were set for facing the actions produced by earthquakes in seismic areas and mentioned in historical manuals (Vivenzio 1783, Palladio 2000, Rondelet J. et al. 2004, Vitruvio Pollione 2008, D'Antonio 2013).

The Romans codified several buildings techniques, like opus pseudoisodomum, opus caementicium, infill wall (emplecton), opus reticolatum with infill, opus incertum, opus mixtum (Vitruvio Pollione 2008). These building methodologies were diffused in their domains but in seismic areas the constructive rules were improved. Timber elements, for example, were a common reinforcement technique for improving the mechanical behaviour of the buildings to horizontal forces, as visible in some archaeological remains in Ercolano (Fig. 1). Moving to modern age, manuals containing several indications to improve the dynamic response of traditional masonry buildings appeared in 18th and 19th century in central and south Italy. In 1783, after a seismic event that destroyed a large part of South-Italy, the manual "Historia de Tremuoti" written by Vivenzio presents antiseismic building typologies, like palaces, reporting plans, sections and prospects with the indications of special reinforcing solutions mainly constituted by timber frames and timber linkage between vertical and horizontal structures. The development in certain areas, like Calabria, of original constructive techniques is the consequence of the characteristics of a land seriously afflicted by important earthquakes in the past. The antiseismic typology known as "casa Baraccata" (Fig. 2) is an example of these strategic measures for facing the seismic problem (Donatelli 2016).

After the 2009 L'Aquila earthquake, D'Antonio (2013) issued a very complete study on the

Figure 1. A Roman House in Ercolano realized with "opus craticium".

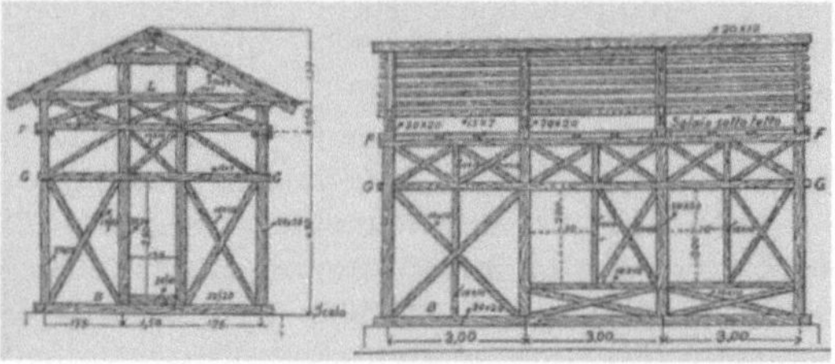

Figure 2. The timber frame system in a project for a "casa baraccata" proposed by Vivenzio in 1783 (from D'Antonio 2013, p. 76).

antiseismic defences observed in the historical buildings of Abruzzi region. The author presents the common measures applied for the reconstruction of the buildings destroyed by previous earthquakes, emphasizing the evolution of some constructive techniques according to the safety demand requested by a territory subjected to seismic actions.

2.2 *The Italian laws for cultural heritage protection*

Among the laws concerning the protection of cultural heritage, Italy has since 2004 a single law gathering the different rules about the safeguarding of historical buildings, the multiple recommendations dealing with the operative procedures defining the conservation project and the future maintenance operations (D.L. 29/10/1999 n.490). Trying to draw out the evolution of Italian laws, codified experiences, presented in manuals above all, turned into national standards in different periods and in 2004 they were gathered into a National Code for Cultural Heritage.

In detail, the second section of the National Code introduces the concept of prevention, emphasizing its importance in the conservation perspective: The conservation of the Cultural Heritage is guaranteed through a coherent, coordinated and planned activity of study, prevention, maintenance and restoration.

- Prevention: set of activities suitable for facing common risks connected to environmental exposure.
- Maintenance: activities for the monitoring of the state of conservation of Cultural Heritage and for keeping its integrity, functional efficiency and identity.
- Restoration: the intervention on the Cultural Heritage finalized to the preservation of its material integrity, its recovery, the transmission of its cultural values to the future.

The National Code introduces also one reference to the seismic risk (article 29), indicating the possibility to adopt the structural improvement intervention for those historical buildings placed in seismic areas. This unique indication in the code, underlines that the problem of the safety of historical buildings is demanded to other specific legislative acts: above all, the standards (mostly guidelines) set by special teams of experts working on this topic.

Looking at the development of the measures set for retrofitting and strengthening of buildings located in seismic areas, we can observe that the role of prevention policies becomes important only during the emergency periods, after seismic events. Since the 1970s, legislative measures were adopted for managing the emergency after each single seismic event. During the 1980s and the 1990s new approaches were proposed in academic field. In this period, the legislative acts were more orientated to reinforced concrete (r.c.) reinforcement technologies, whilst the analysis promoted by different universities introduced new observations to the strengthening criteria (Donatelli A. 2016). Research centers, like academic laboratories or other national research poles, studied the mechanical effects of seismic events on real scale models, simulating common traditional masonry buildings repaired with different strengthening technologies (Magenes & Lagomarsino 2009). As a result, a deep critic to the lack of knowledge towards historical building technologies and their substitution with new reinforced concrete elements was addressed to the national technical board of the Ministry for Public works. Analyzing the seismic events occurred in Italy during the 20th century until nowadays, a relationship between earthquakes and the development of the main legislative acts concerning risk prevention can be draw out:

- In 1908 the seismic risk mapping of the territory was started, after the earthquake that destroyed Messina and Reggio Calabria.
- In 1968, the Belice Valley in Sicily was stroke by a 6.1 magnitude earthquake and in 1974 a complete legislative tool against seismic risk was set by the Ministry for Public Works, providing the technical measures for new constructions in seismic areas and for reinforcing existing buildings.
- In 1976, the 6.4 magnitude earthquake demolished several towns in Friuli and four years later, in 1980, an earthquake by 6.9 magnitude hit the Irpinia area in Calabria. Since 1984, new criteria for the seismic classification of the Italian territory were introduced, looking to new international experiences, like the Japanese and the American ones.
- In 1997, after the Umbria earthquake, with a magnitude of 5.9, the new results coming from the seismic microzonation were used as preventive analysis for post-earthquake interventions.
- In 2002, a 5.8 magnitude earthquake hit Molise region and in 2003 the new national seismic map of the Italian territory (OPCM 3274/2003) was introduced (Fig. 3).

In 2008, after the review of the dangerousness level of the Italian territory, a new comprehensive code of building constructions was set. The 2008 national building code (named NTC2008) is a complete guide for the realization of new buildings and the study of the built heritage in seismic areas.

Since the 1970s, standards for seismic prevention measures were produced by the Ministry of Public Works, following the trends of the construction industry. At the same time, important studies matured between research centres on the historical buildings, but the Ministry of Cultural Heritage did not collect the results in a specific code. In this occasion, in 2009, specific preventive measures for historical buildings were set by the Ministry of Cultural Heritage (Circolare n.617, 2. 2009), following the indications for historical buildings contained in the NTC2008.

According to the National Building Codes (NTC2008), historical buildings and modern ones are subjected to the same preventive measures (D.M. 14.01.2008). In this perspective, seismic microzonation is the fundamental step for any

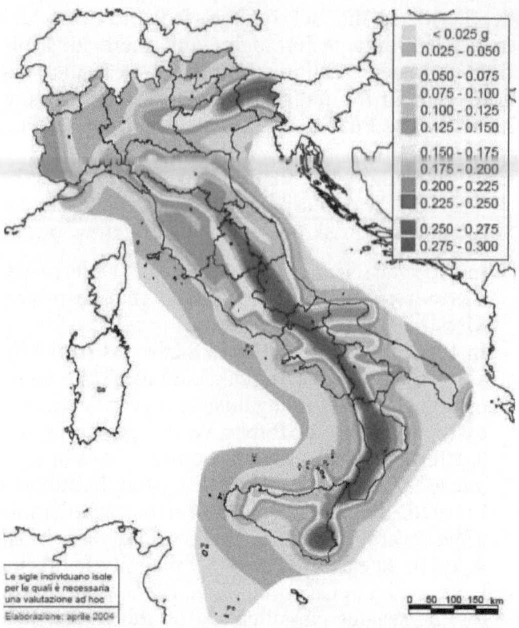

Figure 3. Italian map of the seismic risk. (from the website: http://zonesismiche.mi.ingv.it/mappa_ps_apr04/italia.html).

further decision. If the geological and morphological characteristics of the ground allow maintaining the buildings, technical indications for the strengthening of the structures are given in guidelines and recommendations.

3 THE PREVENTION POLICY OF ITALIAN CODES

3.1 *The early studies on seismic prevention*

Among the main studies on the relationship between earthquake and conservation, important contributions came from the interexchange of contents promoted by different disciplines.

Geology and Earth Physics established new criteria for the design of new buildings and their safety verification, implementing the Seismic Risk Map, introducing the evaluation of the shaking danger level index and promoting the seismic micro-zonation.

Architecture and Structural Engineering gave important contributes for evaluating the vulnerability of the historical buildings (Doglioni 1994 and Magenes 2009). In detail, looking to previous important studies, the following aspects were mentioned as fundamental steps for the analysis:

1. historical analysis through direct and indirect sources;
2. building techniques analysis through diagnostic tests;
3. macro-elements interpretation as basic comprehension of the structural model for future analysis;
4. vulnerability evaluation by the cross relation analysis between the previous points.

The effects of earthquakes drove to new analysis for vulnerability interpretation, like the analysis based on the sub-categories forming the building from a structural point of view. The macro-elements (Fig. 4) are the constitutive parts of the building and can be divided into two-dimensional elements (walls, floors, arches, etc.) and three-dimensional elements (vaults, domes, etc.). The connections between these parts guarantee the structural effectiveness of the building and its response during seismic events.

Considering the past experiences, important properties of the building were taken into account by the architectonic survey analysis. The structural performance of masonry structures can be estimated providing the definition of the following factors:

- the geometry of the building and its structures,
- the characteristics of the masonry texture (single or multiple leaf walls, connection between the leaves),
- the physical, chemical and mechanical characteristics of the components (bricks, stones, mortar);
- the characteristics of masonry as a composite material.

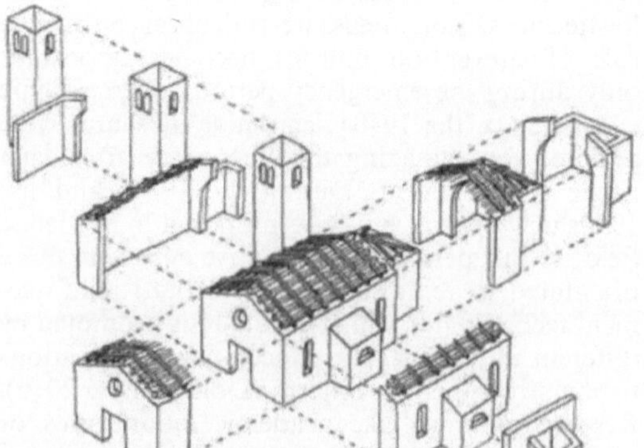

Figure 4. Example of macroelements identification of a church (from Donatelli A. 2016, p. 121).

Also the historical analysis provided a fundamental support to the new study approach. According to the characteristics of the territory, local building technologies were implemented in order to improve the resistance of the structures. In some cases, codified instructions for facing the earthquakes were available for local builders.

Compared to the standardised building techniques, the Ercolano timber frame inserted into masonry walls represents an innovative building solution proposed in an area subjected to seismic activity. The antiseismic techniques used in the past were studied by Giuffrè. This author carried out in the early 1990s the first experimental and analytical studies about the mechanical behaviour of the stonework masonry typologies based on the recognition of peculiar characteristics, named "rules of art", after visual inspection, survey and typological classification (Giuffrè 1993, 1998).

Following the approach indicated by Giuffrè, new studies on the masonry section typologies in different Italian regions were carried out. The first database on historical masonry typologies, providing information on materials properties and wall sections stratigraphy were provided by the GNDT project at the beginning of 21st century. GNDT (National Group Against Earthquake) promoted the study of the components distribution in the masonry sections and the research on the connections between them. The proposed classification was based on: chemical, physical and mechanical analysis.

At the beginning of 21st century, The GNDT project was concluded. The further projects developed by INGV—National Institute for Geophysics and Volcanology, with large public investments in the earth science field, continued facing the problems of building vulnerability. The new projects introduced the following still on-going studies:

- the monitoring of geophysical phenomena in both the solid and fluid components of the Earth;
- 24-hour countrywide seismic surveillance;
- 24 hours real-time volcanic monitoring, early warning and forecast activities;
- Continuous flow of observations from the State-of-the-art networks of geophysical sensors.

3.2 *Recent experiences on seismic prevention*

Thanks to the deep knowledge of the masonry typologies, further studies on its mechanical properties were developed. The Academic Laboratories Network of seismic engineering (ReLUIS), founded on 2003, is a interuniversity consortium with the purpose to coordinate the university laboratories activity on seismic engineering, giving scientific, organizational, technical and financial supports to associated academic institutions and promoting their participation to scientific and technological orientated activities in seismic engineering area, according with national and international research plans.

The activities promoted by the project can be summarized in the following points:

- masonry quality evaluation;
- masonry mechanical properties;
- analysis of the masonry buildings seismic response;
- vulnerability of the masonry buildings at territorial scale
- reinforcing techniques for masonry buildings.

Important results were achieved in the definite systematization of the qualitative interpretation of the failure mechanisms for two main historical building typologies: churches and palaces (G.LA. BE.C 2006). In addition, the researhers experienced the reliability of new software for the numerical model analysis applied to masonry structures and r. c. elements.

The activity experienced by ReLUIS project offered also an important evaluation of the results achieved in the preventive measures proposed in the second half of the 20th century. According to this evaluation, after observing the failure of some interventions carried out on the so called minor built heritage, the measures addressed to seismic risk by the Ministry for Cultural Heritage changed and considered the indications coming from different disciplines. The large number of failures observed after the 1997 Umbria and Marche earthquake showed the vulnerability induced by retrofitting interventions. Retrofitting was intended as a solution for correcting the structural behaviour of masonry structures, classified as elements unable to present a monolithic resource under seismic actions. The intervention was deeply invasive and presented several technical difficulties for connecting new r.c. based elements into masonry buildings. The common solution was a diffused substitution of wooden roofs and floors with stiffer concrete structures. This methodology was even promoted into the recommendations for historical buildings strengthening of the Cultural Heritage Ministry.

In parallel, the antiseismic improvement, a less invasive strengthening methodology, was applied in few cases, through a set of strategies addressed to a diffused connection reinforced between the macro-elements forming the building and localized interventions affecting the masonry quality, like the deep repointing of mortar joints with mixtures presenting a high compatibility with the masonry materials. Examining the collapse mechanisms of the buildings destroyed after recent seismic event, two main factors were identified:

- the low masonry quality of the buildings;
- the lack of compatibility between the existing structures and some retrofitting technical solutions.

After this critical reflection on positive and negative results observed in retrofitted buildings, some research lines later developed by the ReLUIS project (Liberatore & Spera, 2001) studied the potentialities offered by improvement interventions. In detail, after the L'Aquila earthquake, with its important historical centre deeply damaged, the buildings were studied observing also the presence of peculiar technical solutions able to mitigate the seismic effects (Lagomarsino & Podestà 2004a, b). In this sense, the in depth knowledge of the structural characteristics became an important confidence factor for the evaluation of the building safety (Modena 2012).

In recent seismic events, buildings reinforced through new stiff horizontal diaphragms, connected to vertical masonry walls with concrete beams, showed more extensive damages than not retrofitted buildings (Fig. 5 and Fig. 6). The study on the reasons of the observed failures drove to a complete review of the strengthening techniques applied at the end of the last century. The mechanical compatibility between the building elements was assumed as the guiding factor for calibrating future interventions together with an increasing demand on the definition of the mechanical properties of the masonry components. The premises for the future comprehensive building code are defined and will have a predominant impact on further Italian standards for seismic risk prevention.

Figure 5. Effects of a stiff reinforced concrete roof inducing the collapse of the masonry walls after 6th April 2009 earthquake in L'Aquila area (Coppito). (picture by the author).

Figure 6. The partial masonry walls collapse induced by modern stiff horizontal diaphragms after the 2016 Marche earthquake. (Picture by the author).

3.3 *The comprehensive text for buildings*

The large amount of experiences matured in the field of seismic prevention were finally collected in the comprehensive text for buildings, called NTC2008 in Italian language or Comprehensive Building Code.

The Building Code is a performance standard (rather than prescriptive) and its settings where used also for evaluation and mitigation of seismic risk to cultural heritage. The Guidelines set by the Ministry for Cultural Heritage provides the indications for the identification of the final interventions according to the level of knowledge concerning the building characteristics. Considering a wired panorama, in Europe the idea of a study methodology based on different levels of in depth analysis was set in the Eurocode 8 since 2004. In the third part, the Eurocode 8 introduces three different levels of knowledge, accessible through diagnostic investigations, historical analysis and geometrical survey. This approach will be assumed in the Italian Code NTC2008 and used for reaching a confidence index referred to the existing buildings. This approach was even previously mentioned in the guidelines provided by an international commissions belonging to ICOMOS, named ISCARSAH (International Scientific Committee for Analysis and Restoration of Structures of Architectural Heritage). In 2003, the document "Recommendations for the Analysis, Conservation and Structural Restoration of Architectural Heritage" issued by ISCARSAH provided important indications for professionals involved in the conservation practice. This guideline focused on the preservation of historical structures, following the idea of interconnecting different information based on historical, geometrical and mechanical analysis. In Italy, NTC2008 and further application instructions (DPCM 26.02.2011) introduced the so-called "path to knowledge", a procedure for studying the building characteristics from different point of views (taking into account various scientific approaches) based on three main levels of knowledge. They are defined as followed:

1. Limited knowledge level;
2. Relevant knowledge level;
3. Accurate knowledge level.

The first level is based on visual inspections, damage evaluation and qualitative models for the safety verification. This first knowledge level can be used for generic management plans on the building.

When geometric survey and the historical analysis are available, together with some diagnostic characterization, the second knowledge level is obtained. It allows assessing the macro-elements evaluation for setting local interventions on single parts. The accurate level of knowledge is reached when the information obtained by the cross-relation between the mentioned analysis are able to define a complete seismic response evaluation in order to set interventions on the entire construction.

The role of the knowledge path remarks the importance of a high confidence level that each designer should obtain on the historical buildings before the intervention proposal.

4 COSTS OF THE SEISMIC VULNERABILITY AND LACK OF INVESTMENTS IN PREVENTION POLICIES

Taking into account the costs for post-seismic reconstruction and preventive measures (Table 1) provided by the Ministry of Public works (Di Giacomo 2014), the main investments for seismic risk mitigation, it is evident that they are authorized after earthquake destructions. In this sense, the money is not used for a periodical control and improvement of building systems against seismic risks, but for a complete reconstruction of the damaged centers. Maintenance and preventive interventions, in Italy, are not frequently carried out and the lack of control on historical structures does not allow a correct evaluation of the building vulnerability.

Table 1. Earthquake reconstruction costs.

Earthquake	Year	Intervention period	Authorized costs (€ mln)
Valle del Belice	1968	1968–2008	9179
Friuli V. G.	1976	1976–2006	18540
Irpinia	1980	1980–2023	32026
Marche-Umbria	1997	1997–2024	13463
Molise-Puglia	2002	2002–2023	1400
Abruzzo	2009	2009–2029	13700
Emilia	2012	2012–	13300
Total amount			121608

At the beginning of the 21st century, the "fascicolo del fabbricato" (the building booklet) project was proposed for a general overview of the Italian built heritage, in order to catalogue the state of conservation of the buildings placed in active seismic areas. The building booklet never became mandatory in the national codes, but only some regional initiatives mentioned its role for the safety control of the buildings at local contexts. After the recent seismic events, the idea of a booklet containing the information regarding the buildings, its original project and the further transformations became again actual and new proposals for introducing this instrument in the safety evaluation of the built heritage were expressed by different Professional associations (Linee Guida 2017). Despite the large political and professional accord on this idea, the building booklet was not assumed as a practical tool for collecting information on existing buildings. The main oppositions came from the real estate owners, worried about the costs deriving from a negative response that the information collected in the booklet may drive to. As mentioned by Della Torre (2016) in a recent editorial, in Italy the lessons coming from past-experiences gave place to a complete and articulated regulatory framework. Moving from theory to practice, the chronic lack of economic resources for the built heritage is always a limit for the application of the recommendations for seismic risk mitigation: money is mainly used for hiding the cracks in the plaster rather than to operate the necessary analysis for designing proper strengthening interventions on masonry walls (Della Torre 2016).

The large amount of collapsed buildings and damages on the built heritage, after the 2016 earthquake that hit important historical centers like Norcia, in Umbria, represents the occasion for evaluating the set of preventive interventions carried out using the funding allocated after the 1997 earthquake. After the Assisi earthquake, in 1997, Norcia was partially damaged, and the historical center became the field of application of different strengthening interventions. On 30 October 2016, during the morning, a 6.5 magnitude earthquake stroke Norcia causing the collapse of many historical churches.

The destruction of the cathedral in the main square of Norcia had a large public echo, but observing the civil buildings of the historical centers, it can be stated that that they did not collapse and reported, in some cases, limited damages. They can present even large crack patterns, but the interventions carried out after the previous seismic events showed a positive response: no victims and no structural collapses in many civil buildings. A public database for the antiseismic interventions carried out in Norcia is not available (in this case the building booklet could have been a precious tool). According to generic indications provided by the local Engineers Association, the retrofitting of the historical buildings was based on mortar injections and the introduction of new tie rods for increasing the connections between the building structures. These interventions can provide the requested compatibility with the existing structures, following the indications of the most recent studies.

On the other hand, the lack of connections seemed to characterize the monumental buildings, like the S. Benedetto cathedral. According to some observations published after the earthquake (Borri 2016) and a personal qualitative interpretation of the collapse mechanisms (Fig. 7), the church was characterized by a high nave, without shear walls, and a massive bell tower connected to a lateral load

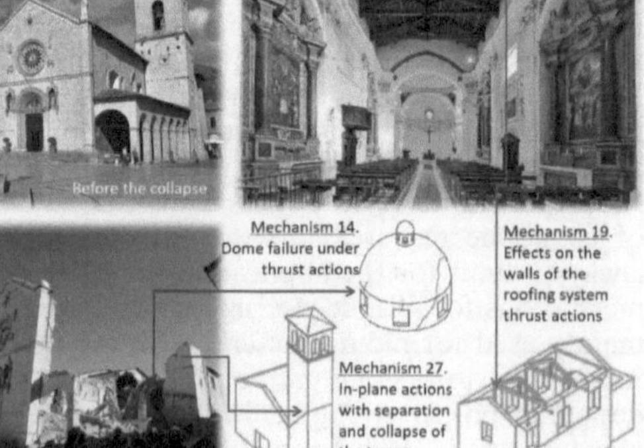

Figure 7. Interpretation by the collapse mechanisms of the structural vulnerability of S. Benedetto cathedral in Norcia (layout issued by the author).

bearing wall: three different collapse kinematics could be assumed. In this case the final response showed by the S. Benedetto cathedral indicates that the protection of large monumental complex represents a difficulty that the local superintendence officers were not able to face. Experts, also coming from academia, commented the case of Norcia as a virtuous example of a correct preventive policy. Nevertheless, critics arrived to the cultural heritage protection offices, institutions unable to evaluate the real vulnerability of the historical churches and more focused to an intervention methodology addressed to the conservation of the architectural surfaces and not to the correct interpretation of the structural mechanical properties of the building.

5 CONCLUSIONS

This work focused on the preventive actions set in Italy and collected in guidelines during a long seismic risk experience. The last series pf earthquakes stroking Lazio, Umbria and Marche in Central Italy remarked the vulnerability of monumental structures (churches above all) and minor historical buildings, usually composed by complex aggregates. For common masonry buildings, the quality of the constructive techniques can represent the turning point between positive or negative response to seismic actions. Even low quality buildings, if founded on soils with specific characteristics, have more chances to mitigate the seismic risk.

The academic experience matured through multidisciplinary programs allows an extended experimentation of the evaluation procedures set by the most recent guidelines for the mitigation of the seismic risk. The prevention measures are now used for public strategic buildings and for important sites, like museums, but the indications are rarely applied to diffused architectural heritage.

The "continuous care" by monitoring on the improved or retrofitted buildings should guarantee the reliability of the interventions, but inadequate strengthening techniques applied without a deep study of the buildings characteristics represent a serious risk for the risk preservation referred to cultural heritage.

REFERENCES

Borri A. (2016). Conoscenza—prevenzione—sicurezza. Qualche ulteriore riflessione. In *Ingenio*, n. 49.

Circolare n.617, 2. (2009). Istruzioni per l'applicazione delle Nuove norme tecniche per le costruzioni. Ministero per i beni e le attività culturali, Roma: *Gazzetta Ufficiale* n.47 26-2-2009 – Supplemento n.27.

D'Antonio M. (2013). *Ita terraemotus damna impedire. Note sulle tecniche antisismiche storiche in Abruzzo*. CARSA Edizioni, Pescara.

Della Torre, S. (2016). Terremoto e prevenzione. In '*ΑΝΑΓΚΗ 79*, Altralinea Editrice, 5–8.

D.L. 29/10/1999 n. 490. *Testo unico delle disposizioni legislative in materia di beni culturali e ambientali*. Gazzetta Ufficiale n. 302, il 27 dicembre 1999

D.M. 14.01.2008, Nuove norme tecniche per le costruzioni. *Gazzetta Ufficiale* n.29, 4.02.2008, Ministero delle Infrastrutture, dell'Interno e Dipartimento Protezione Civile, Roma.

DPCM 26.02.2011, Linee Guida per la valutazione e riduzione del rischio sismico del patrimonio culturale con riferimento alle norme tecniche per le costruzioni. *Gazzetta Ufficiale* n.47 26.02. 2011, Roma.

Di Giacomo, M. (2014). I costi dei terremoti in Italia, Roma, Centro Studi Consiglio Nazionale degli Ingegneri.

Doglioni, F.M. (1994). "Le chiese e il terremoto. Dalla vulnerabilità constatata nel terremoto del Friuli al miglioramento antisismico nel restauro, verso una politica di prevenzione". Lint editoriale, Trieste.

Donatelli A. (2016). *Terremoto e architettura storica. Prevenire l'emergenza*. Gangemi Editore.

EN1998 2004. *Eurocode 8: Design of structures for earthquake resistance.*

Giuffrè A. (1993). *Sicurezza e conservazione dei centri storici. Il caso di Ortigia*. Bari.

Giuffrè A. (1998). *Letture sulla meccanica delle murature storiche*, Kappa ed.

G.LA.BE.C. (2006), Gruppo di Lavoro per la Salvaguardia e la Prevenzione dei Beni Culturali dai Rischi Naturali, Scheda per il rilievo del danno ai beni culturali—Chiese, http://www.protezionecivile.it/cms/attach/adc.pdf, 2006.

Lagomarsino, S. & Podestà, S. (2004a). Seismic vulnerability of ancient churches: part 1. Damage assessment and emergency planning. *Earthquake Spectra*, 20(2): 377–394.

Lagomarsino, S. & Podestà, S. (2004b). Seismic vulnerability of ancient churches: part 2. Statistical analysis of surveyed data and methods for risk analysis. *Earthquake Spectra*, 20(2): 395–412.

Liberatore, D. & Spera, G. (2001). *Il progetto di conservazione: linee metodologiche per le analisi preliminari, l'intervento, il controllo di efficacia*. Firenze, Alinea Editrice.

Linee Guida n. 3 del 1° febbraio (2017). *Il fascicolo del fabbricato per una cultura della prevenzione e della sicurezza integrata*. Consiglio Nazionale dei Periti Industriali e dei Periti Industriali Laureati: CNPI.

Magenes G. & Lagomarsino S. (2009). Evaluation and reduction of the vulnerability of masonry buildings, *Progetto di Ricerca Reluis DPC (2005–2008) – Linea 1, Final report*, Naples.

Modena C. (2012). Conservazione del patrimonio Architettonico e sicurezza strutturale in zona sismica. Insegnamenti dalle recenti esperienze italiane. In *Materiali e Strutture—Nuova Serie*, n. 1–2, 2012 – Donatella Fiorani ed.

OPCM 3274/2003 (G.U. 8-05-2003, n.108) *Criteri generali per la classificazione sismica del territorio nazionale e normative tecniche per le costruzioni in zona sismica.*

Palladio A. (2000). *I quattro libri dell'architettura*, Milano, Hoepli.

Rondelet, J. & Felici, M. (2004). *Trattato teorico e pratico dell'arte di edificare di Giovanni Rondelet*, Roma, Dedalo.

Vitruvio Pollione M. (2008). *De Architectura*, Ed. Studio Tesi.

Vivenzio G. (1783). *Istoria de' Tremuoti*, Napoli.

Innovative Built Heritage Models – Van Balen & Vandesande (Eds)
© 2018 Taylor & Francis Group, London, ISBN 978-1-138-49861-7

Knowledge-based representations applied to built cultural heritage

O. Zalamea, J. Van Orshoven & T. Steenberghen
Spatial Application Division Leuven (SADL), KU Leuven University, Leuven, Belgium

ABSTRACT: Knowledge-based representations provide information in a form that can be used by computer systems. This paper reports the process of finding existing knowledge-based representations for the Built Cultural Heritage (BCH) domain. Strengths and weaknesses of the knowledge-based representations for supporting the development of a system for preventive conservation of BCH are explored. In the BCH-field, multiple actors from various disciplines use a variety of data sources and data models, resulting in a complex data integration problem. In our research, the focus is on explicitly representing 3D features by using the City Geography Markup Language (CityGML) standard. However, CityGML lacks the semantic refinement required for the representation of thematic concepts of the BCH-domain, e.g. vulnerability or legal protection issues; or BCH-components such as pilasters, cornices, etc. To address these shortcomings we present an ontological approach rooted in two pre-existing ontologies: the Geneva-CityGML ontology (an ontological approach based on CityGML), and MONDIS (the monument damage ontology).

1 INTRODUCTION AND PROBLEM STATEMENT

A correct management of Built Cultural Heritage (BCH) assets requires the manipulation of big amounts of information. Traditional information systems created to support the BCH-management have some limitations that we believe can be mitigated by the application of ontological approaches. The following sections describe the characteristics of BCH-assets, how information is managed by traditional information systems, and the research objective and hypothesis of this paper.

1.1 *BCH characteristics*

Built cultural heritage is commonly referred to as the "immovable man-made resources (monuments, group of buildings and sites) which the contemporary generation resolves has a traditional value" (Koboldt 1997). Applied to historic towns and urban areas these resources include mainly urban structures, architectural elements, the landscapes within and around the towns, archaeological remains, panoramas, skylines, view-lines and landmark sites (ICOMOS 2011).

Heritage management is recognized to be a multidisciplinary undertaking, which can benefit from other disciplines and sectors such as management of the natural environment, business, tourism, and finance (Chung 2007). Consequently, BCH-management requires the collaboration of several sectors and actors such as owners, dwellers, local authorities, heritage managers, environmental agencies, etc.

The involvement of multiple actors from various disciplines using a variety of data sources and data models results in a complex data integration problem. The challenge is to develop an approach that facilitates efficient communication and exchange.

1.2 *Traditional information systems*

Several computerized tools are developed for heritage information management (Chung 2007, Bello 2013, Crlislo *et al.* 2014). These tools use relational databases for data organisation and storage which implies some limitations when sharing and integrating data from multiple sources. In this paper, three main limitations of traditional information systems are addressed: i) modification of data models is difficult and implies a considerable amount of work, ii) harmonization of heterogeneous heritage information is hardly possible, and iii) the information that can be retrieved from relational databases is limited.

Modification of current data models is difficult because the use of a traditional systems implies an understanding of the structure according to which the information is stored. When this structure changes the system should change as well, e.g. processes, queries and interfaces have to be redesigned to fit the new structure. This implies a considerable amount of work, especially for systems that are constantly evolving as is the case in the BCH domain (Bluestone 2000).

Data models designed to accommodate information from several sources require users to share exactly the same definition of a common concept. In domains such as BCH it is difficult to share the same terminology and structure, due to the involvement of multiple and multidisciplinary stakeholders (Doerr 2009).

The information that can be queried from a relational database is limited for two reasons: first because it implies prior knowledge of the database structure and second because anything that is not literally stored in the database is not reachable even though this would be very intuitive for the stakeholders. For instance, a historical building located in a protected area should be protected as well. However, if this relationship is not explicitly stored in the database, this deduction cannot be made (De Virgilio *et al.* 2013).

1.3 *The role of ontologies*

Ontological models are developed to address the limitations of traditional information systems. The most common definition (Domingue *et al.* 2011) (Guarino *et al.* 2009) of ontology is: "A formal, explicit specification of a shared conceptualization" (Studer *et al.* 1998). It is formal since data is represented by standardized triplets (subject, predicate, object). For instance, the fact that a historical building has a heritage value is represented by the triplet (historicalBuilding, hasValue, heritageValue). An ontology is explicit since all the terms and properties are formally defined as shown in Figure 1 where not just a historical building is defined but also the property 'hasValue' through its type, domain and range.

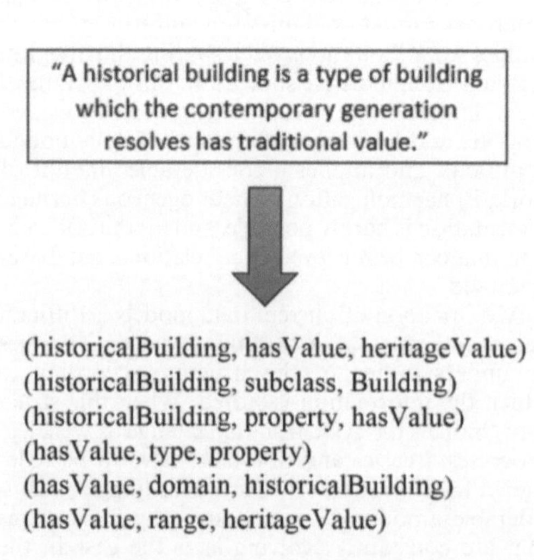

Figure 1. Formal representation of information.

The terminology used to refer to ontologies is broad. Among the practitioners in the field there is not a clear division between what is referred to as "vocabulary" and "ontology". The trend is to use the term "ontology" for more complex, and possibly quite formal collection of terms (World Wide Web Consortium—W3C 2011). Taxonomies being classification schemes arranged in a hierarchical structure are considered as a basic starting point for the construction of an ontology. In this paper, the following concepts are adopted: A taxonomy is a hierarchical structure of classes representing terms of interest of a particular domain. A vocabulary is created when axioms representing relationships among the terms are added to a taxonomy. An ontology is created when logical rules allowing complex logical inferences are added to a vocabulary.

1.4 *Benefits of Ontologies for preventive conservation of BCH*

Semantic interoperability is achieved when computer systems are able to exchange data with unambiguous, shared meaning (Heflin and Hendler 2000). By facilitating semantical interoperability ontologies have the potential to contribute to overcoming the limitations described in section 1.2 and support the aims of preventive conservation of BCH. Preventive conservation of BCH suggests conservation actions considering not only the assessment of state but also periodic assessments of risks and threats. It is meant to facilitate early damage detection by addressing the deterioration causes so that intervention can be kept to a minimum (Forster and Kayan 2009). Semantic interoperability allows data integration between the organizations dealing with BCH-assets and more available data can lead to a better risk assessment. The integrated data can also be used by professionals and scholars increasing the involvement of the community. Section 4 discusses in more depth the benefits of ontological approaches in three use cases related to BCH-management.

1.5 *Objective and hypothesis*

The main objective of this paper is to identify an efficient way to integrate information from various sources to support management of BCH. Our hypothesis is that an ontology can facilitate the mitigation of the limitations previously described. In the following section we define what we mean by 'ontology' and we describe the On-To-Knowledge methodology which we propose for eventually achieving the stated objective. We illustrate the applicability and application of the methodology using a study case related to the World Heritage

City of Cuenca located in the republic of Ecuador, which builds upon (Zalamea *et al.* 2015). In the concluding section we discuss our findings and give an outline of our further work.

2 METHODOLOGY

There are several available methodologies for the construction of ontologies (Casellas 2011). The On-To-Knowledge methodology has been chosen, due its explicit and well elaborated definitions. Moreover, this methodology makes explicit provision for iterative and incremental characteristics allowing improvement of existing ontologies.

2.1 *The On-To-Knowledge methodology*

The On-To-Knowledge Methodology (Staab and Studer 2009) consists of two components: the Knowledge Meta process and the Knowledge process. The Knowledge Meta process leads to the construction of the ontology while the Knowledge process specifies the activities necessary to populate and test the ontology. The methodology is iterative and incremental, allowing a basic ontology to be refined and iteratively improved in any subsequent phase of the process.

2.1.1 *The knowledge meta process*

The Knowledge Meta process consists of five steps: Feasibility study, Kickoff, Refinement, Evaluation, and Application & Evolution. The first three steps are explained in the context of the Cuenca case study while the two remaining steps are addressed in the Discussion section.

2.1.1.1 Feasibility study

The feasibility study identifies the problem, the opportunity areas and the potential solutions. The historical centre of Cuenca is managed by three major actors: the *Cultural Heritage National Institute* (INPC), the *Municipality of Cuenca* and the *World Heritage City Preservation Management* research project at the University of Cuenca (VlirCPM). These actors maintain their own information systems with heterogeneous and overlapping information regarding the BCH in Cuenca. In order to integrate the information the VlirCPM project developed a CityGML-based data model through a dedicated Application Domain Extension (ADE) (Heras *et al.* 2011). The CityGML-ADE model (Fig. 2) provides a useful spatial representation of the BCH-domain. The multiscale application from regions to single buildings and the different levels of detail of data representation from single surfaces to detailed representation of buildings interiors make the CityGML standard a

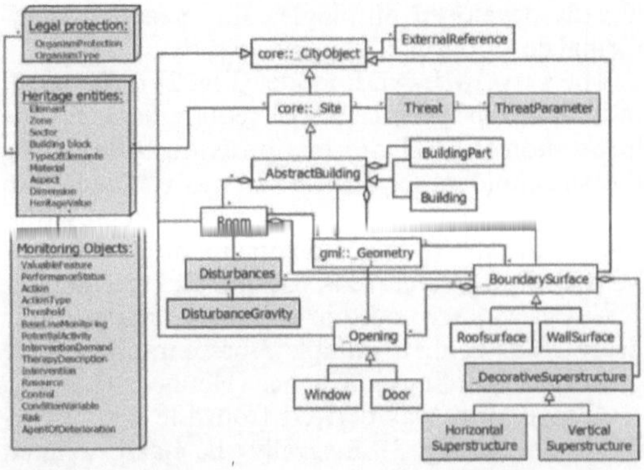

Figure 2. Summary of the CityGML-ADE model for preventive conservation of BCH as developed by (Heras et al. 2011).

suitable solution for the case of Cuenca-Ecuador. The model is meant to be implemented using object-relational database systems which have the limitations identified in previous sections.

An ontological approach could be a potential solution to mitigate these limitations. First, the data structure used by ontologies is a triplet. When modifying data models, triplets are added or removed but the basic data structure '(subject, predicate, object)' does not change. This feature makes modifications of data models less difficult and less time-consuming. Second, the information coming from different actors can be integrated by adding triplets linking data in the different systems. E.g. both the municipality of Cuenca and INPC collect information about building damages but with different level of detail. While the municipality records data about damages for the building as a whole, the INCP records them at the level of the main components of the building. Triplets indicating the relationship between a building and its components can be added making both systems compatible. Third, ontologies promote the use of inferences allowing the discovery of more related information. E.g. when a traditional system is queried about the buildings with a heritage value, the buildings without stated heritage value will not be considered in the response. On the contrary, in an ontological approach according to the formal definition in Figure 1 all the historical buildings have a heritage value. Thus, all the buildings will be part of the response since by definition historical buildings are the ones that have a heritage value even though such value might not be assigned yet.

2.1.1.2 Kickoff

The Kickoff step of the On-To-Knowledge ontology captures requirement specifications, looks for already developed ontologies and creates a semi-formal description of the ontology.

The CityGML-ADE model (Fig. 2) is the result of a thorough specification of requirements for the integration of BCH information as available among the stakeholders in Cuenca. It was derived from several interviews and workshops with representatives of the INPC, the Municipality of Cuenca, the VlirCPM project, citizens, tourists and researchers. It was previously tested in the study area and was accepted as being a suitable representation of the Historical Buildings in Cuenca (Heras *et al.* 2013).

A taxonomy was derived from the CityGML-ADE model (Fig. 3). Basically, the classes related to the preventive conservation present in the ADE were mapped as taxonomy classes and the generalization associations were converted in ontology classes and subclasses. For this purpose we used Protégé, a free and open source suite of tools to construct domain models and knowledge-based applications with ontologies (Stanford Center for Biomedical Informatics Research 2015).

Next a search for knowledge-based representations of the BCH taxonomy terms was performed. Literature, catalogues and search engines were used to perform the search. Catalogs and search engines recommended by the World Wide Web Consortium (W3C) were used in this research: Linked Open Vocabularies (Atemezing *et al.* 2013), Swoogle (Finin *et al.* 2004) and Falcons (Cheng *et al.* 2008). The search was concluded with the selection of 22 knowledge-based representations (Upper table in Fig. 4).

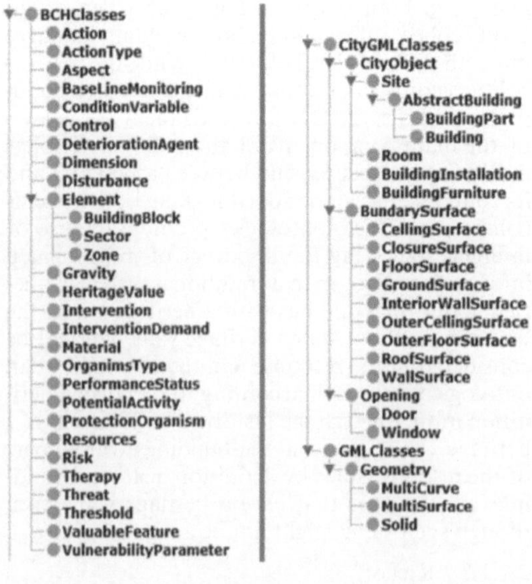

Figure 3. Taxonomical representation of the CityGML-ADE model.

bag	geneva	lgdo	owl
crm	gndo	lode	prext
dbpedia	gov	log	schema
dogont	juso	mtlo	vaem
ecrm	juso.kr	mondis	
frbrer	km4c	oslo	

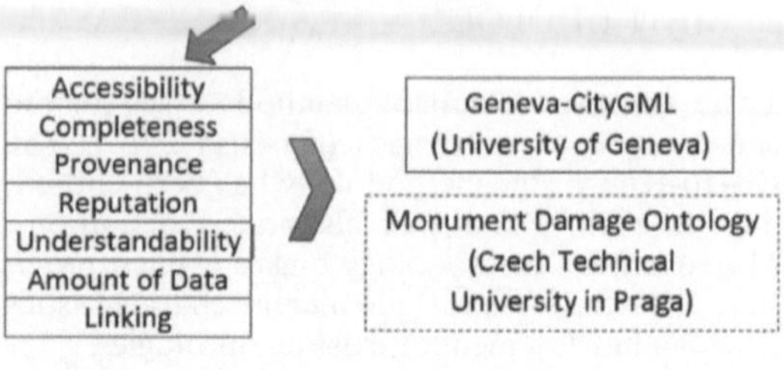

Figure 4. Evaluation of the knowledge representation candidates.

2.1.1.3 Refinement

In the refinement step the knowledge-based representations found in the previous step were evaluated. The first criterion addressed was availability. For BCH we verified if the ontology was downloadable. If the ontology was available then the degree of completeness was computed so that priority could be given to the most complete ones. Then provenance and reputation of the creator of the ontology was verified to assess the ontology's usability. Understandability is another important criterion ensuring that the knowledge-based representation can be interpreted by both machines and humans. Finally, additional criteria such as the amount of data, the number of incoming and outgoing links were added in case several ontologies were found to be equivalent.

As a result of the evaluation activity of candidate BCH-ontologies in the refinement step, two ontologies were selected (Fig. 4): The CityGML ontology developed by the University of Geneva in Switzerland and the Monument Damage Ontology developed in the framework of the MONDIS research project. The University of Geneva CityGML ontology can be accessed through http://cui.unige.ch/isi/onto/citygml2.0.owl. It provides a three dimensional physical representation of buildings and their components. It was the only one based on the CityGML standard which makes it the only possible solution meeting the user requirements. The Monument Damage Ontology supports the documentation and analysis of monument damages. It is available at: http://kbss.felk.cvut.cz/ontologies/2011/monument-damage-taxonomies.owl.

Figure 5 shows in solid lines the classes covered by the Geneva-CityGML and with dash lines the ones represented in Mondis. The: Aspect, Dimensions, Heritage values, Valuable features, etc. are

Figure 5. Terms in the taxonomical representation of the CityGML-ADE-model (Figure 3) covered by the selected ontologies.

concepts not covered by any of the two selected ontologies. The creation of such concepts is under way.

3 RESULTS

In this section we provide a general description of each of the two selected ontologies.

3.1 *The Geneva-CityGML ontology*

The ontological version of the CityGML standard developed by the University of Geneva (Métral and Cutting-Decelle 2011) provides means to represent different city elements such as buildings, tunnels, bridges, waterbodies, etc. Within the scope of this research only the building module was used (Fig. 6). The city objects can be represented with more or less specifications depending on the Level Of Detail (LOD) that is required.

LOD0 represents a building as a set of 2.5-dimensional surfaces: for example one surface for the footprint of the building and other for the roof edge. LOD1 represents the outer shell of the building as a vertical extrusion of the footprint, resulting in a block. LOD2 allows the representation of architectural details like roof overhangs, columns, or antennas. The class used to represent these details is the 'BuildingInstallation' and the

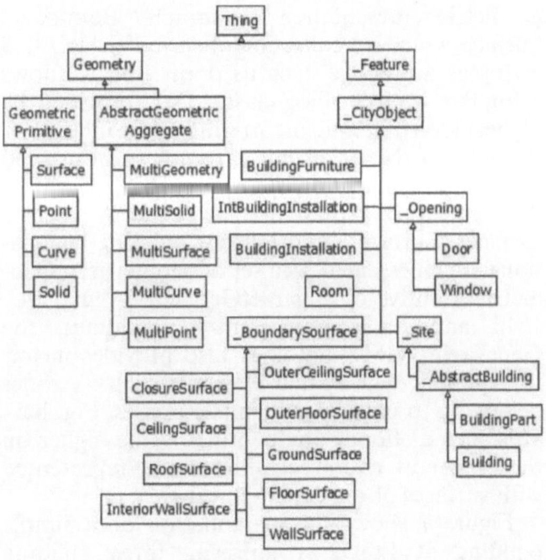

Figure 6. Main classes in the Geneva-CityGML ontology.

```
<owl:NamedIndividual rdf:about="&bcho;0108007012">
  <rdf:type rdf:resource="&bldg;Building"/>
  <bldg:consistsOfBuildingPart rdf:resource="&bcho;groundPart"/>
  <bldg:consistsOfBuildingPart rdf:resource="&bcho;wallPart"/>
  <bldg:consistsOfBuildingPart rdf:resource="&bcho;roofPart"/>
</owl:NamedIndividual>
<owl:NamedIndividual rdf:about="&bcho;wallPart">
  <rdf:type rdf:resource="&bldg;BuildingPart"/>
  <bldg:outerBuildingInstallation rdf:resource="&bcho;wallInstallation"/>
</owl:NamedIndividual>
<owl:NamedIndividual rdf:about="&bcho;wallInstallation">
  <rdf:type rdf:resource="&bldg;BuildingInstallation"/>
  <bldg:boundedBy rdf:resource="&bcho;wallMultiSurface"/>
</owl:NamedIndividual>
<owl:NamedIndividual rdf:about="&bcho;wallSurface">
  <rdf:type rdf:resource="&bldg;WallSurface"/>
  <bldg:lod2MultiSurface rdf:resource="&bcho;wallMultiSurface"/>
</owl:NamedIndividual>
<owl:NamedIndividual rdf:about="&bcho;wallMultiSurface">
  <rdf:type rdf:resource="&gml;MultiSurface"/>
  <gml:surfaceMember rdf:resource="&bcho;wall1Polygon"/>
</owl:NamedIndividual>
<owl:NamedIndividual rdf:about="&bcho;wall1Polygon">
  <rdf:type rdf:resource="&gml;Polygon"/>
  <gen:value rdf:datatype="&www;XMLSchema#string">
        <gml:LinearRing>
              <gml:pos>104.00 32.00 0.00</gml:pos>
              <gml:pos>76.00 16.00 0.00</gml:pos>
              <gml:pos>76.00 16.00 15.00</gml:pos>
              <gml:pos>104.00 32.00 15.00</gml:pos>
              <gml:pos>104.00 32.00 0.00</gml:pos>
        </gml:LinearRing>
    </gen:value>
</owl:NamedIndividual>
```

Figure 7. Simple building instatiation using Geneva-CityGML.

geometric representation used for it is '_Boundary-Surface' which is a collection of surfaces. At LOD3 openings are added such as doors and windows using the 'MultiSurface' class. LOD4 provides the highest level of detail, incorporating the interior of the building. New elements as rooms, furniture and interior building installations can be represented. A room can be represented as a solid or a set of surfaces. Furniture and interior building installations are represented by a set of geometries: 'GeometricPrimitive' or 'AbstractGeometricAggregate'.

In addition to representing buildings, the Geneva-CityGML ontology also provides means to represent new spatial elements such as zones belonging to particular thematic classes. E.g. hazards such as floods are identified to be higher on the border of rivers; location can be represented with surfaces of the geographical area.

Figure 7 shows the instantiation of a simple building at LOD2, comprising three building parts: ground floor, walls and roof.

Building parts as walls are represented by the 'BuildingInstallation' class. Building installations are 'bounded by' multisurfaces which are composed by polygons. In the same fashion roofs and ground surfaces can be represented.

Figure 8 shows the physical representation of building represented in Figure 7.

3.2 *The Mondis ontology*

The Mondis ontology (Fig. 9) focusses in 4) the representation of damages present on buildings. Its classes are grouped in four thematic clusters: component, event, damages and risk assessment.

The cluster components allow to assign physical and functional characteristics to the cultural heritage objects. Complex and normal objects can be described by defining respectively multiple or single components. The classes 'Use', 'Structural-Type' and 'FunctionalType' provide further details concerning the purpose for which the object was created and the current occupation. The classes

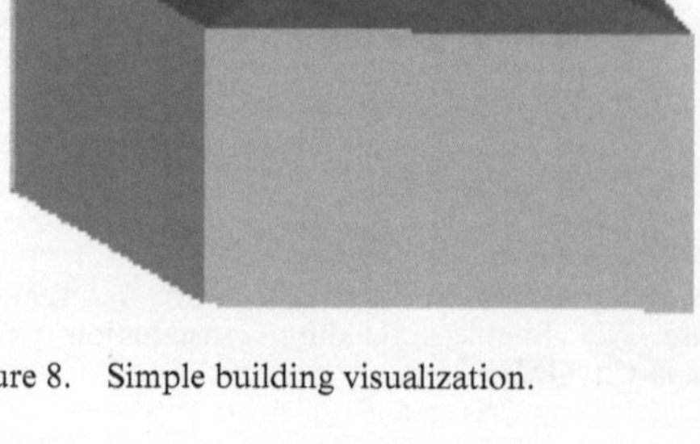

Figure 8. Simple building visualization.

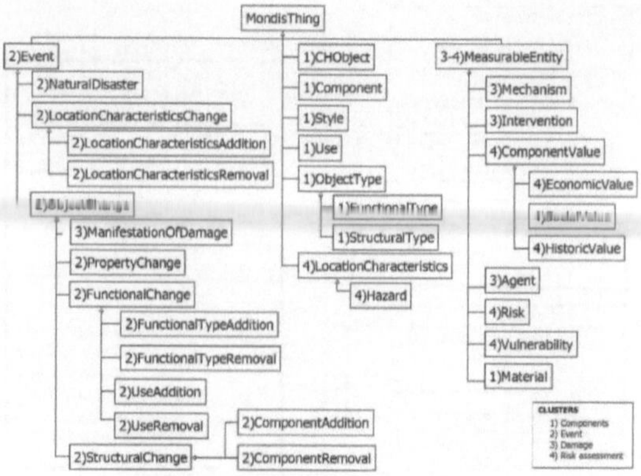

Figure 9. Main classes in the Mondis ontology.

'Style' and 'Material' describe the building techniques used in the creation of the object.

The class 'Event' encompasses subclasses as 'NaturalDisaster', 'ObjectChange', and 'LocationCharacteristicsChange'. The 'NaturalDisaster' class is related with the risk assessment classes. The class 'ObjectChange' includes changes on function, use, and structure of components. External changes affecting the heritage object are documented using the 'LocationCharacteristicsChange' class.

Damages are produced by mechanisms triggered by an agent. Interventions in the heritage object repair or mitigate the damage.

The risk assessment classes describe risks and vulnerabilities of heritage objects. Heritage objects are vulnerable against hazards which jeopardize the value of the components in the heritage objects.

The following example shows how a 'detachment' in one of the walls of the building represented in Figure 7 is encoded by the Mondis ontology.

4 DISCUSSION AND FUTURE WORK

4.1 *The BCH-ontology*

An alternative for CityGML-ADE not used in the evaluation of existing ontologies is the Industry Foundation Classes (IFC) data model. Substituting CityGML-ADE by IFC is however possible and such modification will not infer an excessive amount of work since ontologies are easy adaptable to changes.

Geneva-CityGML and Mondis as the selected ontologies need to be merged for creation of an ontology for BCH-management. This requires the establishment of a number of links between both. The first obvious link to add the 'Building' class from Geneva-CityGML as a subclass of the 'CHObject' class of Mondis. This will allow

```
<owl:NamedIndividual rdf:about="&bcho;0108007012">
  <rdf:type rdf:resource="&mondis;CHObject"/>
  <hasComponent  rdf:resource="& bcho;wall1Polygon"/>
</owl:NamedIndividual>
<owl:NamedIndividual rdf:about="&bcho;wall1Polygon">
  <rdf:type rdf:resource="&mondis;Component"/>
  <hasComponentManifestationOfDamage rdf:resource="&bcho;detachment"/>
</owl:NamedIndividual>
<owl:NamedIndividual rdf:about="&bcho;detachment">
  <rdf:type rdf:resource="&mondis;ComponentManifestationOfDamage"/>
  <hasAgent rdf:resource="&bcho;water"/>
</owl:NamedIndividual>
<owl:NamedIndividual rdf:about="&bcho;water">
  <rdf:type rdf:resource="&mondis;Agent"/>
</owl:NamedIndividual>
```

Figure 10. Example of the encoding of a detachment in one of the walls of a building using the Mondis ontology.

it to inherit Mondis properties as 'hasManifestionOfDamages', 'hasVulnerability', etc.

Second, the other '_cityObject' classes from Geneva-CityGML need to be related as equivalent to the 'Component' class of Mondis. These links will allow for integrated spatial and thematic queries.

In order to provide a complete ontological framework for BCH management, finally classes and properties expressing heritage values, heritage features, aspects and dimensions must be created and linked to the Mondis class 'CHObject'.

4.2 *The BCH-ontology use cases*

What needs to be done in the second component, i.e. the Knowledge process of the On-To-Knowledge methodology, is to define new use cases in order to test the applicability and the universality of the ontology. The aim of these use cases is to confirm the improvements of using ontological approaches over traditional systems hypothesised in the above described feasibility study of the Knowledge metaprocess.

The first use case will deal with the modification of current data models. One important cause of data model modifications in the BCH domain is the switch between methodologies (De Virgilio *et al.* 2013). E,g., methodologies for risk assessment are becoming very popular. This use case will implement two methodologies for risk assessment of historical buildings: the quantitative methodology developed by (Jimenez 2001) for the city of Cuenca-Ecuador, and the quantitative-qualitative methodology developed by (Paolini *et al.* 2012) for the archeological site of Petra-Jordan. Both methodologies will be compared for a traditional system on the one hand and an ontological approach on the other hand.

A second use case will integrate heterogeneous information from different datasets. The information about damages of buildings in Cuenca is maintained by the city municipality and the INCP. The city municipality records damages presented in specific building elements while the INCP records the damages in the building in general. A query about the damages in both datasets is currently not possible since both authorities use different data models, building identifiers don't match, and the scale at which damages are recorded is different. To allow the integrated query of both datasets, the ontological approach will add new relationships for enabling records of the buildings to match and linkage of buildings to their building elements.

The third planned use case involves the generation of information through the use of logical inferences. An example is the derivation of heritage values from descriptive information as the architectural style, building technique, materiality, etc.

5 CONCLUSION

We illustrated the applicability of the Knowledge Meta Process of the On-To-Knowledge methodology to compile an ontology specifically directed to (the preventive conservation and) management of BCH. The methodology consists in identifying the applicable pre-existing ontologies (the University of Geneva CityGML-ontology and the Mondis ontology) and in their combination with a source taxonomy (CityGML-ADE). The Geneva CityGML ontology can be considered to be an ontological equivalent of the CityGML standard which provides a three dimensional physical representation of buildings and their components. The Mondis Ontology supports the documentation and analysis of damages to monuments. Both ontologies provide a good starting point for further expansion since they complement each other. Validation of the BCH-ontology remains to be carried out and will be done in line with the Knowledge Process of the On-To-Knowledge methodology.

REFERENCES

Atemezing, G.A., Vatant, B., Troncy, R. & Vanderbussche, P.-Y. (2013). Harmonizing services for LOD vocabularies: a case study. *In*: *WASABI 2013, Workshop on Semantic Web Enterprise Adoption and Best Practice, Colocated with ISWC 2013 Workshop, October 22, 2013, Sydney, Australia*. Sydney, AUSTRALIA.

Bello, L. (2013). GIS as a tool to improve the management of Vista Alegre district. *Polytechnic Institute José Antonio Echeverría, Architecture Faculty, La Habana, Cuba*, 34, 19–31.

Bluestone, D. (2000). *Challenges for Heritage Conservation and the Role of Research on Values*. Los Angeles: The Getty Conservation Ins titute.
Casellas, N. (2011). Methodologies, Tools and Languages for Ontology Design. *In: Legal Ontology Engineering: Methodologies, Modelling Trends, and the Ontology of Professional Judicial Knowledge*. Springer Netherlands, 57–107.
Cheng, G., Ge, W. & Qu, Y. (2008). Falcons: searching and browsing entities on the semantic web. *In*: *Proceedings of the 17th International Conference on World Wide Web*. Beijing: ACM, New York (2008), 1101–1102.
Chung, Y.S.S. (2007). Heritage Planning. *The SAA archaeological record—The Magazine of the Society for American Archaeology*, 7 (4), 42–46.
Crlisle, P.K., Avramides, I., Dalgity, A. & Myers, D. (2014). The Arches Heritage Inventory and Management System: a standards—based approach to the management of cultural heritage information.
De Virgilio, R., Maccioni, A. & Torlone, R. (2013). Converting Relational to Graph Databases. *In*: *First International Workshop on Graph Data Management Experiences and Systems*. New York, NY, USA: ACM, 1:1–1:6.
Doerr, M. (2009). Ontologies for Cultural Heritage. *In*: S. Staab and R. Studer, eds. *Handbooks on Ontologies*. Springer.
Domingue, J., Fensel, D. & Hendler, J.A. (2011). *Handbook of Semantic Web Technologies*. London New York: Springer-Verlag Berlin Heidelberg.
Finin, T., Ding, L., Pan, R., Joshi, A., Kolar, P., Java, A. & Peng, Y. (2004). Swoogle: Searching for knowledge on the semantic web. *In*: *Proceedings of the thirteenth ACM international conference on Information and knowledge management*. Presented at the Conference on Information and Knowledge Management 2004, New York: ACM New York, NY, USA © 2004, 652–659.
Forster, A. & Kayan, B. (2009). Maintenance for historic buildings: A current perspective, 210–229.
Guarino, N., Oberle, D. & Staab, S. (2009). What is an ontology? *In*: *Handbook on ontologies*. Springer–Verlag.
Heflin, J. & Hendler, J. (2000). Semantic Interoperability on the Web. *In*: *Proceedings of Extreme Markup Languages 2000*. Proceedings of Extreme Markup Languages 2000, 111–120.
Heras, V., Steenberghen, T. & Zalamea, O. (2013). A GIS based tool for a Preventive Conservation Management Approach. *In*: *Reflections on Preventive Conservation, Maintenance and Monitoring of Monuments and Sites—PRECOM3OS UNESCO Chair*. Leuven, BE: ACCO, 86–93.
Heras, V., Wijffels, A., Cardoso, F., Vandesande, A., Santana, M., Van Orshoven, J., Steenberghen, T. & Van Balen, K. (2011). A value-based monitoring system to support heritage conservation planning. *Journal of Cultural Heritage Management and Sustainable Development*, 3 (2), 130–147.
ICOMOS. (2011). "The Valletta Principles for the Safeguarding and Management of Historic Cities, Towns and Urban Areas". Adopted by the 17th ICOMOS General Assembly, November 28.
Jimenez, J.C. (2001). Vulnerabilidad Sismica de las Edificaciones de la Ciudad de Cuenca mediante técnicas de simulación. Master en Ingenieria Estructural. Escuela Politecnica Nacional, Quito.
Koboldt, C. (1997). Optimizing the Use of Cultural Heritage. *In*: M. Hutter and I. Rizzo, eds. *Economic Perspectives on Cultural Heritage*. London: Palgrave Macmillan UK, 50–73.
Métral, C. & Cutting-Decelle, A.F. (2011). Ontologies for Interconnecting Urban Models. *In*: *Ontologies in Urban Development Projects*. Springer London, 105–122.
Paolini, A., Vafadari, A., Cesaro, G., Quintero, M.S., Van Balen, K. & Pinilla, O.V. (2012). *Risk Management at Heritage Sites: A Case Study of the Petra World Heritage Site*. UNESCO.
Staab, S. & Studer, R. (Eds.) (2009). *Handbook on ontologies*. 2nd ed. Springer–Verlag.
Stanford Center for Biomedical Informatics Research. (2015). Protege [online]. Available from: http://protege.stanford.edu/about.php [Accessed 26 Nov 2015].
Studer, R., Benjamins, V.R. & Fensel, D. (1998). Knowledge engineering: Principles and methods. *Data & Knowledge Engineering*, 25 (1–2), 161–197.
World Wide Web Consortium—W3C. (2011). W3C Semantic Web Activity [online]. Available from: http://www.w3.org/2001/sw/ [Accessed 19 Aug 2014].
Zalamea, O., Steenberghen, T. & Van Orshoven, J. (2015). From a CityGML to an ontology-based approach to support preventive conservation of built cultural heritage. Presented at the 19th AGILE International Conference on Geographic Information Science, Finland: Submitted on 08/12/2015.

Innovative Built Heritage Models – Van Balen & Vandesande (Eds)
© 2018 Taylor & Francis Group, London, ISBN 978-1-138-49861-7

Digital tools supporting conservation and management of built cultural heritage

S. Della Torre, R. Moioli & A. Pili
ABC Department, Politecnico di Milano, Italy

ABSTRACT: The paper aims at comparing some experiences, in which conservation plans have been introduced, to test the implementation of different knowledge management tools. The discussion refers to Italian context. Italian Cultural Heritage Code enforced in 2004, and Guidelines published by Lombardy Region in 2003 and issued in 2005, which define contents and methodology for maintenance plans and scientific reports. The comparison focuses and costs and benefits of the adoption of simple or sophisticated digital tools, such as GIS or relational databases. The ultimate question is if the choice of the tool may influence the owners' willingness to implement continuous care procedures entailing not only the needed intervention works, but knowledge management as well.

1 KNOWLEDGE MANAGEMENT IN PLANNED CONSERVATION: ALTERNATIVE TOOLS AND IMPLEMENTATION ISSUES

The purpose of this paper is to compare some experiences, in which conservation plans have been introduced, to test the implementation of different knowledge management tools.

The discussion refers to Italian context. Italian Law, in the Cultural Heritage and Landscape Code enforced in 2004, states that "Cultural Heritage conservation is guarantee through coherent, coordinated and planned activities of study, prevention, maintenance and restoration".

The Preventive and Planned Conservation methodology, which complies with this regulatory article, is based on scheduling and financial planning of activities aimed at protecting, preserving and enhancing Cultural Heritage.

In 2003, Lombardy Region published the "Guidelines for the Conservation Plan and scientific final balance", then officially issued in 2005, which define contents and methodology (Della Torre 2003/a).

The ultimate question is if the adoption of different tools could influence the owners' willingness to implement continuous care procedures entailing knowledge management (Dann 2004).

2 THE CONSERVATION PLAN: PHASES AND STRUCTURE

According to Lombardy Region Guidelines, the Conservation Plan is the implementation tool that allows the coordination of the phases and activities of this process, responding to the needs of Cultural Heritage management over time in accordance with the systemic and dynamic approach distinguishing the Preventive and Planned Conservation.

A Conservation Plan encompasses four documents: Technical Handbook, Conservation Program, Economic Budget and User Handbook.

The Technical Handbook defines the framework of knowledge and previous events concerning the building, through the description of materials and construction techniques, decays, activities and technical operations, diagnostics, relationship between the elements, problems and uncertainties to which the element is exposed in relation to intrinsic (materials and construction techniques) and extrinsic (context and use) features.

The attempt to introduce some kind of risk assessment is a feature that is worthy to be underlined, as it marks a difference compared to the structure of ordinary maintenance plans. As focused on the conservation of authenticity in historic buildings, on the line followed by the definitions of Prevention and Maintenance given by the Italian law in 2004, Lombardy Guidelines trigger inspection and conservation assessment instead than routine maintenance works, but furthermore ask to investigate uses and building components, to detect their fragility due to intimate singularities, which could develop decay processes if border conditions change (Della Torre 2003/b). This understanding of hidden issues is very important as it leads to improve use and management as leverages to prevent changes in the context, which may activate mechanism cause-effects leading to damages. It means a special form of predictive approach in preventive conservation, pretty customized on individual characteristics of single objects. In this case the analogy with preventive medicine (Van Balen 2014) goes forward, as primary prevention gets customized thanks to deep investigation of individuals, as happens in the so called "personalized medicine" (Brigham 2010). Also the difference with traditional predictive and planned maintenance requires to be clarified: the evaluation of latent incompatibilities leads to control the border conditions, and has nothing to do with service-life statistic predictions that lead to optimize the timing of substitutions.

The Conservation Program is connected to the Technical Handbook and defines by whom, when and how the activities have to be carried out. It organizes the data referring to involved professionals, necessary means to fulfilment, expected times and period of execution, and specific operating modes. Subsequently, the activities are programmed and systematized into a time schedule.

The Budget Sheet systematizes the economic resources management through definition and planning of costs related to prevention, control, diagnostics and monitoring, restoration works and maintenance works.

The User Handbook is the document that summarizes in a non-technical language the information contained in the Technical Handbook and in the Conservation Program, targeting end-users in order to pursue conservation through continuous care and proper management. The User Handbook expresses and highlights the active role of the user in the conservative strategies, as a guarantor of good conservation, being he/she who daily lives the building.

The compilation of a Conservation Plan, whatever the digital tool used, is developed through four phases, by activities that in some cases are repeated and intersected.

The first step consists in activities aimed at gathering information and data, at defining the structure of the plan according to the formal characteristics and specific management needs, and at breaking down the building in technological elements, assigning them a univocal code.

The second phase consists in data entry within the documents.

The third one is the implementation phase, which is developed through the issue, execution and closure of Work Orders.

The last one consists in updating the files, by feedback from implementation, and potential reprogramming of the planned actions, depending on the outcomes of the activities and on managerial choices.

The updating overlaps with a new data update phase, thus connoting the cyclic nature of the conservative process and allowing stratification of knowledge.

3 INFORMATION TECHNOLOGY TOOLS FOR CONSERVATION PLAN

The Conservation Plan can be compiled through various Information Technology tools that allow data logging and management. Several researches have been carried out aiming at defining digital tools to manage the big amount of data produced in the conservation process. The main alternative was between GIS-based applications, successfully developed for the documentation of conservation works on surfaces (Capponi et al. 2001; Coccoli et al. 2003; Cacace 2006; Baracchini et al. 2007; Bartolomucci 2008) and relational databases. As oriented to a more comprehensive understanding of the building structure, relational databases proved to be flexible and performing in the different phases, in which data should be filed, updated or retrieved (Canziani & Turati 2007), producing customization to user needs, which is a general tendency in building maintenance management (Chanter & Swallow 2008; Talamo & Bonanomi 2015).

The focus of this paper is therefore no longer on this kind of alternative, but on issues related to implementation of Lombardy Region guidelines, as it has been detected that often owners are not willing to pay also for the costs of the system itself, which they perceive as very high. Therefore, in several cases simpler tools have been implemented, giving up many performances provided by relational databases, using instead simple electronic tools such as Word processing software and spreadsheets.

Undoubtedly both the recording and the consultation/management of the information (especially the latter) vary considerably depending on the adopted tool. Word processing software and spreadsheets certainly have the advantage to be largely used on the market, besides to be widespread also in Open Source and Free versions; but they present difficulties for information upload and management, above all in presence of many data, being static software designed for other purposes, where it is not possible to formulate queries and statistics.

Databases and Information Systems in particular, offer the ability to manage, through queries and statistics, and so in dynamic mode, large amounts of related data, allow attachment associated with data and the management and sharing of the whole system by more users (even with differentiated access levels) through remote server access, and also offer the ability to relate to other Information Systems.

4 CASE STUDIES

Experiences chosen for the comparison test are: Conservation Plan of Santa Maria del Lavello

Monastery in Calolziocorte, 2014, compiled at first with word processing software, while in a second moment data have been systematized in an Information System; Conservation Plan of Masegra Castle in Sondrio, 2015, compiled through spreadsheets; and lastly, Conservation Plan of Royal Villa in Monza, 2013, compiled through the Information System "Planet—Architectural Heritage" was developed within the "Monumentenwacht Italy" project as an evolution specific for the cultural heritage of I.S. "Planet Maintenance" widespread in Italy for Facility Management (Benatti, Borgarino & Della Torre 2014).

Not being able to measure in terms of data storage (for example GBytes) to compare methods that use different recording systems, it is possible to use as a common base the number of pages produced when printing the Plan documents. The Plans, just for the first data recording and excluding the on-going management phase, have produced for Masegra Castle about 400 pages, while for Santa Maria del Lavello Monastery about 200 pages. Concerning the files of photographic documents and cognitive information of various kinds, these have been classified into folders. The amount of data produced in Monza are more consistent: already in the first data recording phase the printing would reach 4,600 pages and, by adding the Inspection Reports, it would exceed 6,000 pages.

Planet presents a data organization which follows a hierarchical structure by levels, from general to detail ones, where each single technological element, even if grouped into clusters, is uniquely defined by an alphanumeric code. Basically, Planet is a system that works per objects.

The layout of the Plan is displayed in the User Interface displayed in two modes: a tree representation mode (showing the hierarchy of the structure and the groups categorization (classes, and technological elements) and in a table format (where the database attributes are visualized).

The implementation and management of the data within the System is done through thematic panels organized by sections according to the documents and phases of the Plan, complying with Lombardy Guidelines. In detail, the panels for compiling, updating and printing the data of the Technical Handbook, the Conservation Program, the Economic Budget and the User Handbook belong to the section "Identity Card (Anagrafica)"; in the "Schedule" section the functions are to be found, which allow processing the Conservation Program in a time schedule on the basis of specific periods; in the "Work Orders" section it is possible to manage the documents related to the executive phase of the Plan.

As structuring the Plan within Planet, it is possible to define the accounts of each stakeholder involved in the conservation process, by defining for each of them the personal data, role and profile type, contacts, authority or company belonging, access permissions and credentials, that is modification permission or consultation only. Due to its special complexity, the management process in Monza case requires further explanation in a devoted section.

5 PROCESS, ACTORS AND MANAGEMENT TOOLS FOR THE ROYAL VILLA OF MONZA

The Royal Villa is one of first cases in Italy of public heritage management with a public/private partnership through the instrument of concession. This instrument can be seen as an innovative model of public works financing that allows the realization of projects with combination of private capitals in support of the public asset (Boniotti & Della Torre 2016).

The applied model provided for a first phase of restoration carried out in the 2012–2014 period (determining what can be named a "zero time" in reference to the conservation status of the building) and management for the next twenty years. The private investor can recover the invested sums through the granted management of the historic building for a defined number of years.

At the end of the concession period, currently in progress, the concessionaire will have to return the Royal Villa to the grantor in the same optimal condition achieved by the restoration works; to this purpose, the contract foresees Preventive and Planned Conservation practices, through a constant control of the execution of maintenance plans, to avoid another restoration. Therefore, Conservation Management has to be integrated in Facility and Property Management, which the concessionaire manages through "Manpro.net" Information System. The two Information Systems have been linked, allowing the coordination of spaces, human resources and both the activities aimed at opening to the public, and those addressed to care and conservation of the historic building and its precious interiors.

The preliminary phase involved the acquisition of site data and subsequent insertion into the S.I. Accurate as-built reports of the huge conservation works just carried out were available. This is therefore an effective application on the volume of data that a restoration normally produces and potentially makes available to the following phases (Lai & Pili 2014). As-built reports were compiled for individual rooms, often grouping the elements, as for some elements the exact location of diffuse interventions was impossible to retrieve, since they were almost homogeneously distributed on all elements due to the advanced degradation the building had reached.

For the purpose of testing the feasibility and practical usefulness of this method of knowledge management, it is important to notice that after the exercise of filling in the relational database the huge amount of data coming from the as-built reports, the tool has been really implemented by the concessionaire, under the control of the grantor. This has been the result of the vision, which from the beginning prompted the project, but it seems also the outcome of the awareness of the management about the economic convenience of a preventive approach to the very expensive conservation of the interiors.

Therefore, it has been already possible to test the system also in the implementation phase. For the inspection activities, an Inspection Sheet was created for each room. The Inspection Sheet consists of a textual part and a graphic part: the first relating to the checks to be carried out for each type of element with the subdivision of activities by professional skills (architect and restorer), the explication of the operative procedures and a space where it is possible to record the inspection results; the second part represented by the printing of the associated DWF file, in which the individual technological elements are identified and encoded.

During the inspections, the results were recorded manually on the Final Inspection Report, while the decay maps were reported on the graphs. The inspection activities clearly were preceded by the issuance of a Work Order which, at the end of the inspection campaign, has been closed. The results were then implemented within the I.S. on the single element level and the Final Inspection Report attached as electronic file.

6 COMPARISON

The experiences performed using various tools have made possible to effectively compare the methods.

In all cases the buildings went through a breakdown into parts from general (building) to detail (element), defining groups according to localization, technological classes and elemental typologies, identifying the single elements by univocal alphanumeric codes.

For the comparison, as evaluation parameters, the following criteria have been adopted:

- practicality in data systematization;
- ease of consultation and management data on the long run;
- storage modes of attached documents;
- capability of shared implementation and consultation between the actors of the process;
- capability of integration with graphic bases and other management systems and information environments;
- costs.

The first point, practicality in data systematization, refers to the possibility to compile data coming to the detail of the single element, without necessarily resorting to grouping in order to avoid overproduction of static files that would be difficult to manage over time. For Masegra Castle and Santa Maria del Lavello Monastery Plans, the Technical Handbook and Conservation Program sheets have been compiled grouping the elements in technological classes, while Planet proved to be scalable from groups to single elements.

As for the data consultation and management case, the criterion is inherent to the fact that even a trivial search seems to be easier and more immediate within an information system than with the other methods, without neglecting the advantages deriving from the data queries provided by statistics and filters service, not supported by static systems.

Concerning attached documents storage, in the case of Planet software, they are uploaded one-time into the database and can be associated with several elements and levels of detail, without necessarily requiring duplication. Thus overloads into storage is avoided, which has occurred in the other cases, where the files were organized within folders saved in static and local mode,. The attachments in I.S. also have a direct link with related element, including DWF files.

With reference to data sharing, the "Planet—Beni Architettonici" manages, by IP address (therefore through network protocols), data recorded on a remote server accessed through authorized profiles, possibly with different access privileges. Each subject in the process participates in the management of the Plan within its roles and functions. The step from server to cloud logic is obvious.

Another important aspect consists in interaction and integration. In the last few years, various ways of compiling plans have been tested, including procedures for association of data to a two-dimensional and three-dimensional graphical base, also parametric through links to GIS and BIM systems (Pili 2014; Osello & Rinaudo 2016). The interoperability and/or integration with other Information Systems, e.g. conceived for administrative or management issues (Facility and Property Management), has already been successfully tested even involving parametric models of the building (Talamo & Bonanomi 2015). The interaction and integration between the Systems may even allow to conceive Planet platform as a Common Date Environment, a function that lacks in the other methods exposed.

Last, but not less important, is the parameter concerning costs. The criterion is twofold: there is a cost for software and hardware, and a cost in terms of human resources and time consumed for the setup and the population of the databases.

Software for text editing and spreadsheets are widely spread on the market, also Open Source and Free versions, while Planet is a licensed I.S. This would benefit for the first two. On the other hand, costs would increase considerably during the information management using static files.

It is clear that the initial investment for the purchase of the Information System is going to be amortized over time, with considerable savings in the phase of global process management.

Population costs are the most critical problem observed during the project: actually, a lot of time has been spent by skilled operators in analysing available information and filing selected data into the databases, whatever the kind of tool adopted.

7 CONCLUSIONS

It is clear that information and data are increasing considerably over time, so static computing tools such as text editors and spreadsheets are unable to respond to the dynamism of conservative process. Instead, Relational Databases and Integrated Systems enable to handle rationally and sustainably these large amounts of data and documents. An Information System/Relational Database (as for instance "Planet—Beni Architettonici") allows a dynamic-systemic management of the whole process and associated data, with the possibility of interaction and integration with other Facility and Property Management Information Systems, using network protocols for sharing information.

The most interesting perspective concerns integration (because of affinity in architectue and in the logic of shared information management) with modern BIM technologies, already implemented on Built Cultural Architectural Heritage (Historic-BIM). The main advantage would come from interoperability with the other documents produced in the various phases, and especially in projects and diagnostics report. Interoperability would allow an automatic or semi-automatic population of the database, or even better management activities and projects could refer to a Common Data Environment, in which data would be filed once forever and could be retrieved without ambiguity (Della Torre 2017).

In other words, innovation is expected to lower costs and to make convenient and acceptable for owners an effective knowledge management oriented to continuous care of heritage buildings.

REFERENCES

Baracchini C. et al. (2007). Verso un sistema unico di riferimento per la documentazione di restauro: storia e sviluppi di SICAR w/b. In *e-Arcom07. Sistemi informativi per l'architettura*, Atti del convegno (Ancona, 17–19 maggio 2007): 84–89. Firenze: Alinea.

Bartolomucci, C. (2008). I GIS per la conservazione dell'architettura storica. *MondoGIS*, 66/67: 13–15

Benatti, E., Borgarino, M.P. & Della Torre, S. (2014). Planet Beni Architettonici. Uno strumento per la Conservazione Programmata del Patrimonio storico—architettonico. In Stefano Della Torre (ed.), *ICT per il miglioramento del processo conservativo, Atti della Conferenza internazionale Preventive and Planned Conservation, Monza—Mantova, 5–9 maggio 2014*: 13–29. Firenze: Nardini Editore.

Boniotti, C. & Della Torre, S. (2016). Innovative funding and management models for the conservation and valorisation of public built cultural heritage. In Biscontin, G. & Driussi G. (eds.), *Eresia e ortodossia nel restauro, Progetti e realizzazione, Atti del Convegno di Studi, Bressanone 28 giugno – 1 luglio 2016.*: 105–114. Venezia: Arcadia Ricerche.

Brigham, K. (2010). Predictive Health: The Imminent Revolution in Health Care. *Journal of the American Geriatrics Society*, Volume 58, Issue s2 (State-of-the-Art in Longitudinal Studies on Aging): S298–S302.

Cacace, C. (2006). GIS: uno strumento per la rappresentazione organizzata delle informazioni sui processi di restauro e conservazione. *Bollettino ICR*, n.s., 13: 19–26.

Canziani, A. & Turati, F. 2007. Obblighi conservativi nella gestione di un patrimonio architettonico. Analisi delle criticità e proposte di strutturazione delle informazioni di supporto. In *Third International Conference on Maintenance and Facility Management. Proceedings*: 17–24. Rome: CNIM.

Capponi G. et al. (2001). Un'applicazione nella mappatura dei materiali costitutivi e dello stato di degrado della Torre di Pisa. *Usi e consumi dell'informazione geografica*, Atti della 3° Conferenza MondoGis (Roma, 23–25 maggio 2001): 229–236. Roma: MondoGis.

Chanter & Swallow (2008). *Building Maintenance Management*. Wiley.

Coccoli, C. Catellani. M., Cavagnini, G. & Treccani, G.P. (2003). G.I.S. e conservazione programmata: un caso applicativo. Il piano di manutenzione della Parrocchiale di Vilminore di Scalve (BG), *MondoGIS*, 37: 22–25.

Dann, N. (2004). Owners' Attitudes to Maintenance, *Context*, 83: 14–16.

Della Torre, S. (ed.) (2003/a). *La conservazione programmata del patrimonio storico—architettonico: linee guida per il piano di manutenzione e il consuntivo scientifico*. Milano: Guerini.

Della Torre, S. (2003/b). Riflessioni sul principio di compatibilità: verso una gestione dell'incompatibilità. In *Dalla Reversibilità alla compatibilità*, atti del convegno di Conegliano Veneto (13–14 giugno 2003): 27–32. Firenze: Nardini Editore.

Della Torre, S. (2017). Un bilancio del progetto BHIMM. In Della Torre, S. ed., *Built Heritage Information Modelling/Management—BHIMM*, online (www.ingenio-web.it).

Lui, A. & Pili, A. (2014). Attività propedeutiche alla redazione del Piano di Conservazione della Villa Reale di Monza. In Fanzini, D. & Moioli, R. (eds.), *Il Master in conservazione preventiva e programmata per la valorizzazione dei beni culturali del Distretto Le Regge dei Gonzaga*: 33–38. Firenze: Nardini Editore.

Osello, A. & Rinaudo, F. (2016). Cultural Heritage Management Tools: The Role of GIS and BIM. In Stylianidis, E. & Remondino, F. (eds), *3D recording, documentation and management in cultural heritage*. Dunbeath, Caithness: Whittles Publishing.

Pili, A. (2014). La gestione del Piano di Conservazione con software GIS e BIM: metodologie a confronto. In Biscontin, G. & Driussi G. (eds.), *Quale sostenibilità per il restauro?, Atti del Convegno di Studi, Bressanone, 1-4 luglio 2014*: 713–722. Venezia: Arcadia Ricerche.

Talamo, C. & Bonanomi, M. (2015). *Knowledge Management and Information Tools for Building Maintenance and Facility Management*. Berlin: Springer.

Van Balen, K. (2014). Gaining Insight into Preventive Conservation. Lessons Learnt from Medicine and from ongoing Exchanges. In Della Torre. S. (ed.), *La strategia della Conservazione programmata. Dalla progettazione delle attività alla valutazione degli impatti, Atti della Conferenza internazionale Preventive and Planned Conservation, Monza—Mantova, 5–9 maggio 2014*: 11–20. Firenze: Nardini Editore.

Innovative Built Heritage Models – Van Balen & Vandesande (Eds)
© 2018 Taylor & Francis Group, London, ISBN 978-1-138-49861-7

The role of Monumentenwacht: 40 years theory and practice in The Netherlands

H. Heinemann & S. Naldini
Faculty of Architecture and the Built Environment, Delft University of Technology, Delft, The Netherlands

ABSTRACT: Monumentenwacht organisations are appreciated for their supporting role in preventive conservation. In The Netherlands, Monumentenwacht is active for more than 40 years and has been integrated in conservation practice. In this paper, an analysis of its *modus operandi* is presented, based on a qualitative multiple case study research. The underlying research question is how Monumentenwacht has influenced daily maintenance practice, and to identify its role in conservation practice. The findings were compared to the theoretical preventive conservation cycle of signalling, diagnosis, treatment, and documentation. The research indicated a paradigm shift from Monumentenwacht's role of signalling problems and monitoring to being the main support for maintenance, in particular for private owners and for owners of non-iconic buildings. As influencing factors changes within the Dutch heritage care were identified, such as a decrease of support for diagnosis and treatment by conservation specialists, and an increase of owners directly approaching contractors.

1 INTRODUCTION

The concept of implementing preventive conservation by offering owners of monuments accessible and affordable support to carry out maintenance has been disseminated throughout several European countries within the last decades. Its origins lies in the Dutch Monumentenwacht, which was founded as a non-profit organisation in 1973 with the intention to mediate between the government which funded large scale restoration campaigns and owners who were responsible to maintain their buildings, yet lacked knowledge to do so (Vos, 2010). As the creation of Monumentenwacht addressed a practical problem, its set-up was a bottom-up and client-orientated approach, and has remained as such. Parallel to the dissemination of the Monumentenwacht concept (Federatie Monumentenwacht Nederland, 2002), its role within preventive conservation has been discussed increasingly within academia (e.g., Vandesande and Van Balen, 2016). According to Van Balen (UNESCO PRECOM³OS Chair, 2016) Monumentenwacht's role is comparable to the health care concept of "*secondary prevention: means of monitoring that allow an early detection of the symptoms of the unwanted effects (damage).*" This view aligns with Monumentenwacht's own definition of its role, namely to preserve "*authentic materials and finishes…, the increase of restauration cycle from the custom 30–50 years to 50–80 years … and increasing awareness of owners by implementing planned maintenance*" (Vereniging van Provinciale Monumentenwachten Nederland, 2015).

With more than 40 years of activity, Monumentenwacht has been widely accepted and integrated in Dutch heritage care practice. This allows a review of Monumentenwacht's role during time, including how it has influenced owners on implementing preventive maintenance. The expected results can indicate whether Monumentenwacht's practice and related theory align, and which factors changed questioning possible deviations. This paper makes part of a threefold investigation related to Monumentenwacht North Brabant: a preliminary research on the importance of Monumentenwacht, focusing on conservation specialists and the context within Dutch conservation practice (Van Hees et al., 2015, Naldini et al., 2017); this paper on Monumentenwacht's *modus operandi,* and will be completed with a publication based on a survey among members of Monumentenwacht about the utilisation of Monumentenwacht's report and owner's decision making (Heinemann and Naldini, 2017).

2 METHODOLOGY

The underlying research question was how Monumentenwacht has influenced daily maintenance practice, and to identify its role in conservation practice. The findings were compared to the theoretical preventive conservation cycle of signalling,

Table 1. Overview of the case studies.

#	Listing*	Original function	Present function	Owner	Century
1[†]	R	Church	Church/events venue	Church council	15th
2[†]	R	Church	Church	Church council	13th
3[†]	R	Church	Events venue	Private	20th
4[†]	R	Church	Church	Church council	12th
5[†]	R/–	Factory	Multifunctional, vacant	Province	20th
6	R	Military	Atelier	Municipality	17th
7	–	Housing	Housing	Private	19th
8[†]	R	Housing	Commercial	Trust	18th
9[†]	–	Housing	Housing	Private	20th
10	R	Health centre	Housing	Private	20th
11[†]		Housing	Housing	Private	16th
12	G	Agricultural	Agricultural	Private	18th
13	R	Housing	Offices	Foundation	19th
14	R	Bridge	Bridge	Public body	19th
15[†]	R	Town hall	Town hall	Municipality	16th
16	–	Commercial	Commercial	Private	19th

[†]: Case studied in more detail *R: National Monument;
G: Municipal monument; –: not listed

diagnosis, treatment, documentation as mentioned in Vandesande (2017). In order to understand Monumentenwacht's *modus operandi* in practice and its influence on owners and monuments, a qualitative multiple case study was carried out. Due to the confidential nature of the data, the 16 cases are presented anonymously; Table 1 gives an overview of their characteristics. Besides being inspected by Monumentenwacht North Brabant, the selection criterion was that the total of the cases represent a variety of typologies, construction periods, degree of listing, and type of ownership. These characteristics can influence the availability of subsidies and thereby potentially the implementation of recommend works.

As the 10 branches of the Dutch Monumentenwacht have slightly different operational methods, a restriction was made to Monumentenwacht North Brabant in order to eliminate factors due to organisational differences. To understand its *modus operandi* the inspection and reporting methods and its current and previous positions within conservation practice were reviewed. For all cases, the most recent inspection report was analysed regarding its scope, signalled problems, terminology, and recommendations given. 8 cases were studied in more detail (Table 1), including interviews with owners, managers, user, and if present restoration architects and contractors. The open-ended interviews aimed at understanding how the interviewees perceive Monumentenwacht, how the reports were used, and whether recommendations given in the reports were followed. The respective monuments were visually inspected by the authors, involving Master's students of the Faculty of Architecture and the Built Environment, TU Delft (course 'Building Conservation Assessment') in order to identify the strengths and limits of Monumentenwacht's inspection and reporting method. An unexpected obstacle in this research was the limited storage of documents related to previous inspections and maintenance activities (e.g., costs, materials), in particular in cases of non-iconic buildings. Therefore, the available data varied per case.

3 MONUMENTENWACHT NORTH BRABANT

In The Netherlands, Monumentenwacht is organised per province. All branches have in common of being a non-profit organisation with the aim to support owners to carry out maintenance timely by means of monitoring and signalling problems. The services offered vary per province, with some offering additionally long-term maintenance planning, estimation of maintenance costs, and advice on subsidies. The motivation was partially financial, as the subsidy Monumentenwacht receives varies per province, and partially commercial reacting on market demands. This research is limited to cases inspected by Monumentenwacht North Brabant (founded in 1975), as it only services are inspecting and reporting, and thus allowing a better study of its role within the maintenance process. Monumentenwacht North Brabant consciously chose not to extend its services as other branches have done in order not to interfere with the working field of conservation specialists (e.g., restoration architects). In other provinces were such developments had taken place, it has created resentment within the restoration market (Mullink, 2010). The Province of North Brabant, understanding the benefits of an independent Monumentenwacht, grants therefore sufficient subsidy to make a participation in the commercial market unnecessary.

Per year, Monumentenwacht North Brabant alone inspects approx. 3000 objects (Monumentenwacht Noord-Brabant, 2017). Large or fragile objects are inspected yearly, smaller objects such as houses biyearly. This frequency is considered by Monumentenwacht North Brabant as the minimum to prevent decay, to identify poor workmanship and damage timely, and to remain in close contact with the owner, and thereby maintaining a low threshold for owners to ask Monumentenwacht questions about maintenance. As the inspection report is Monumentenwacht's main product, much attention has been given to improve its readability by colour coding the degree of severity and including more pictures of the signalled damage. Furthermore, reports contain an overview drawing of the roof, a short description of the elements and materials inspected, damage symptoms encountered, and occasionally advice is given. The provided advice can range from carrying out straightforward actions such as a regular cleaning of gutters to consult a specialist. During informal talks immediately after the inspection the findings are explained to owners, including how to implement the recommendations and pointing out potential risks when not taking action timely.

4 MAINTENANCE SUBSIDIES

Subsidies are one incentive for owners to carry out preventive maintenance, and influenced the maintenance works of the cases studied. Therefore, a brief overview of the main developments of maintenance subsidies will be given. The Dutch government introduced maintenance subsidies for national monuments in 2006 to stimulate preventive conservation, while slowly decreasing subsidies for restoration (Ministerie van OC en W, 2006). Essential innovations were that subsidised works could be distributed over a period of six years instead of one year and long-term maintenance plans were required, which led to more cost effective planning of works. Depending on the ownership and building typology, 20–65% of the maintenance costs could be subsidised.

When reviewing the initiative (Ministerie van OC en W, 2008), it was noticed that some aspects were not sufficiently considered, such as the fact that preventive maintenance can only postpone replacement of historic material but not supress decay, and hence creating a financial gap for necessary restoration works. With the most recent maintenance subsidy rule, the tendency to focus on restrained maintenance works while excluding restoration has been continued (Ministerie van OC en W, 2016). Noteworthy is the fact that the membership fees and inspection costs for Monumentenwacht can be declared.

5 EXPLORATIVE CASE STUDIES

In the following paragraphs, the 16 cases (Table 1) are discussed in groups based on similarities identified: iconic churches, common national monuments, neglected monuments and privately owned housing.

5.1 *Iconic churches*

Cases 1 and 2 are both iconic churches of high national importance. Due to their national significance and high costs of restoration and maintenance, both churches received additional, incidental subsidies for exceptional large restoration campaigns (BRGR subsidy, Ministerie van OC en W, 2000).

The first case is a 15th century church erected in brick and cladded with different types of natural stone. Since the late 19th century, major conservation campaigns have taken place every 20–30 years. Nowadays, maintenance works are continuous and as part of the subsidy requirements planned for a period of six years. Maintenance and restoration works are coordinated by a restoration architect, and Monumentenwacht has been inspecting the church for several decades. Its role is clearly defined as a supportive instrument to signal problems timely, in particular in remote locations. Due to the significance of the building, possible interventions are discussed with the Dutch Heritage Care Agency and if necessary, experts consulted. The amount of maintenance carried out strongly depends on subsidies received, and in case of limited resources, priorities are made to keep the building watertight.

Similarities exist with the second case, a 13th century cathedral (Fig. 1). Since the 1850s, three major restoration campaigns have been carried out. As the church had been re-consecrated as a catholic church after being protestant for a long period of time, the initial restoration aimed to emphasise the new status as a cathedral, and its poor state of conservation was thus unacceptable. Nowadays, the

Figure 1. Detail of the façade of case 2, showing a diversity of natural stone.

church still requires restoration works due to the size of the object and the natural ageing of materials. This aspect is relevant for the owner, as in future, subsidies will be granted for maintenance, rather than for restoration. Maintenance works are coordinated by a restoration consultant employed by the church and an external restoration architect; Monumentenwacht inspects the church yearly, and is additionally onsite for small maintenance measures. In this case, the craftsmanship of the inspectors and their ability to access remote locations, combined with the low fees are the main advantages by the owner. Monumentenwacht's role is to signal damage; diagnosis and planning of interventions are tasks of a restoration architect, supported by experts and advisors of the Cultural Heritage Agency. For example, natural stone specialists are involved for diagnosis and treatment; a necessity understood by the owner, restoration architect and Monumentenwacht inspectors.

5.2 *Common national monuments*

Conservation of national monuments without a high public exposure is less discussed in public. Examples are the cases 3, a transformed chapel, 4, a small parish church, and 15, a town hall. Together, they represent common maintenance approaches with an increasing dependency on Monumentenwacht.

Case 3, an early 20th century chapel has been transformed like many former religious buildings into an events venue. The building is now privately owned, the financial resources for maintenance are limited, and no subsidies are available. Only Monumentenwacht investigated the building, and its state is according to the reports reasonable-moderate. In particular the slate roof, the lead flashings and the rainwater drainage are in a poor condition, and inevitably the timber roof structure shows rot. Although the owner understands the advantage of carrying out repairs timely, only the most urgent are carried out due to financial restrictions.

A small, 12th century Romanesque church (case 4) has been inspected by Monumentenwacht for over 25 years. Exceptional characteristics are its Roman tuff stone walls dating from the construction phase (Fig. 2), and its timber roof structure, which is one of the eldest still existing in The Netherlands (Fig. 3). Maintenance is managed by a member of the parish, supported by Monumentenwacht and a foundation for preservation of churches helping the owner to apply for subsidies. The motivation to join Monumentenwacht was to have information on necessary maintenance. During the last three decades, changes related to the state of conservation were identified

Figure 2. Case 4. Detail of the wall, showing layering of the tuff stone and detachment of the pointing.

Figure 3. Case 4. The timber roof is one of the eldest in The Netherlands.

the inspection reports. As financial resources, despite receiving maintenance subsidies, were not sufficient to carry out all works timely, intervention were carefully chosen: urgent works such as repairing the roof and gutters to prevent further damage were carried out; less urgent was a repair of the façades. Local deterioration of the tuff stone had already been identified in 1990, yet still considered acceptable. In 2015, the walls had not yet been repaired, and degradation had continued, with increasing layering and spalling (Fig. 2). Hence, Monumentenwacht recommended replacing the historic tuff stone, and subsequently the owner approached a contractor for advice. Damage causes were neither investigated, nor were alternatives to stone replacement discussed. However, due to the significance of the Roman tuff stone, a natural stone specialist from the Cultural Heritage Agency became involved in the preliminary discussions. A more detailed approach, including damage mapping and diagnosis, was suggested due the complexity of degradation mechanisms of tuff stone. This situation represents a common development in Dutch maintenance practice, where owners do not consult specialists, and hence the steps of diagnosis and developing a treatment strategy are omitted, and contractors are directly approached to discuss repair options.

For a municipality owning a 16th century town hall (case 15) Monumentenwacht's role is to support its maintenance department. Being trained for modern buildings, its employees do not have sufficient knowledge of historic buildings in order to identify maintenance needs. The town hall has been inspected for several years by Monumentenwacht, and overall, the building is well maintained. In the report of 2015, mainly problems related to poor workmanship such as of lead flashings between the slated roof and the adjacent brick wall were signalled. For the owner, Monumentenwacht's specific knowledge on historic buildings and materials, and their ability to inspect the roof are appreciated. In addition, there is a long-standing relationship between the inspector and the caretaker, facilitating communication and down-to-earth explanation of findings and possible solutions. Maintenance plans are adapted to the recommendations made by Monumentenwacht. The municipality does not apply for maintenance subsidies, as in their opinion the efforts and costs related to drawing up the documents for subsidies (long-term maintenance plan, costs calculations) are too high as granting of a subsidy is uncertain.

5.3 *Neglect and transformation*

Abandoned or neglected buildings are inspected by Monumentenwacht as well. The underlying motivation of the owner is usually not to implement maintenance but to have a starting point for large-scale restoration works and/or transformation of the building. Inspections are then not a regular monitoring inspection but a baseline measurement, where the state of condition is recorded for the first time.

Case 5, an abandoned 20th century factory complex, is an example where prior to a transformation Monumentenwacht was asked to carry out a baseline measurement. The resulting report was used to apply for subsidies to confirm the poor state of conservation, and to negotiate necessary works with the contractor. However, the owner did not consider implementing regular maintenance as he intended to sell the object once it had been transformed.

The building dating from the 1920s poses two challenges for monitoring and maintenance; it is partially an early reinforced concrete structure with a load bearing brick façade and partially an early reinforced concrete frame with brick infills. Within heritage practice, historic reinforced concrete is a recent working field, also for Monumentenwacht. Secondly, the building has been neglected for more than a decade, leading to severe damage caused by leaking roofs, broken downpipes, vandalism and general lack of maintenance. Consequently, Monumentenwacht identified several poor elements, requiring repair or further research. The problems encountered included corrosion of the reinforced concrete, growth of higher plants on the façades, and cracks in the brick façade (Figures 4 and 5). Despite of a minor cause, which could have been corrected easily on time or even prevented by keeping the vacant building watertight, damage accumulated, requiring now more complex interventions as also load bearing elements have been affected by moisture ingress.

Figure 4. Case 5. Detail of exterior wall with defect downpipe, biological growth and deformation in the brick wall.

The building has been transformed partially into a location for creative industry, while planning a re-use of the entire building for housing and as hotel. Repairs were determined by the availability of tenants and their needs. Due to a limited budget, interventions not related to making the rented zones suitable to let were postponed, such as the repair of the roof and downpipes, and thereby the state of conservation of the upper floors is still declining (Figures 4 and 5). Despite the complexity of the building and the severity of damage, no further investigations have been carried, despite being recommended by Monumentenwacht. The report is considered by the contractor as sufficient to estimate the amount and costs of works, and commence. The visual inspection by students and the authors confirmed the local severity of damage and necessity to consult experts. For example, correlations between cracks in load bearing brick walls were identified which cannot be signalled with Monumentenwacht's element based approach. This case exemplifies common practice where Monumentenwacht reports are misunderstood by owners and practitioners as a full survey, and the risks of not investigating damage causes and of postponing maintenance not seen.

Case 8, a former 18th century single-room house with workshop, was abandoned prior to its restoration in 2002. The object is owned and managed by a professional company investing in heritage, and currently commercially used. Many elements, such as the roof structure, interior finishes, and windows were replaced and are thus recent. The overall state of conservation is good, also based on the fact that most elements are recent. Monumentenwacht's advice mainly relates to improving poor workmanship. As the owner understands the need of maintenance, it was for him self-evident to have the building inspected by Monumentenwacht, appreciating their capability, compared to other surveyors, to carry out minor repairs immediately. The owner speaks regularly with employees of Monumentenwacht about reports and maintenance. However, the reports are considered by the owner as temporary documents which are discarded when new reports are available. The benefits of documenting information and consulting previous reports to understand changes during time are not considered.

Figure 5. Case 5. Interior showing biological growth on the wall behind a defect down pipe, on the ceiling detached plaster.

5.4 *Private owners of housing*

National monuments with a housing function and privately owned form a special case as the financial incentives for maintenance differ. Instead of a subsidy they can only deduce maintenance costs from their yearly tax declaration or apply for a special loan. In addition, the threshold for private owners to consult specialists is higher, and hence they depend stronger on Monumentenwacht.

Case 11, a 15th century mansion privately owned and used as a holiday house, is an example for such a constellation. The building has been inspected by Monumentenwacht yearly, and the overall state of conservation has been evaluated as moderate-poor. The combination of limited resources of the owner and only seasonal occupancy is a reason why not all recommended interventions, even though urgent, have been carried out. As a consequence, the condition of the building has declined during time. For example, a neglected crack running through and through an exterior brick wall with an adjacent broken downpipe caused a leakage resulting in severe decay inside the building. The potential risks of not carrying out further research and not intervening timely are accepted by the owner, as the resources are lacking to do so.

During this research, students inspecting the building confirmed many findings of Monumentenwacht, yet also confirmed the limits of Monumentenwacht reports, which are not always obvious for owners. Similar as in case 5 crack patterns were identified; in this case cracks in the vaulted ceilings and in the brick façades (Van Hees et al., 2015) indicated a structural problem which Monumentenwacht suspected and therefore had recommended a structural survey, which was not carried out.

6 MONUMENTENWACHT'S CHANGING ROLE

In theory, organisations as Monumentenwacht complement preventive conservation practice, supporting the cycle of signalling, diagnosis, treatment and documentation. Monumentenwacht's activities ought to focus in particular on signalling and documenting damage, and communicate its findings to owners and conservation specialists active in the successive steps within the conservation cycle. However, reviewing Monumentenwacht in practice indicated that some owners and professionals do not see it not anymore as a supportive, signalling instrument to implement preventive maintenance.

The cases studied identified several aspects characterising the relationship between owners and Monumentenwacht. On the one hand, there is a good and often long-standing relationship between both parties. The inspectors are familiar with the building, and the changes it underwent during time. For owners, the threshold to ask questions and discuss the reports is low. Monumentenwachters are appreciated for their expertise, and if financially possible, owners carry out the recommended interventions.

Regarding the role of Monumentenwacht within the conservation process, it was seen that for the maintenance of iconic churches (cases 1 and 2) Monumentenwacht is, as initially intended, a supportive instrument to signal problems, which are discussed and solved by a board of experts. The cases 3,4,5,8 and 15, however, revealed an increasing dependency of owners on Monumentenwacht and a decreasing involvement of restoration professionals. Maintenance or restoration specialists are not consulted, and decisions are made by the owner and contractor based on the inspection report. The only control mechanism is Monumentenwacht when evaluating the workmanship afterwards. The justification to rely on the inspection reports only was not always limited financial resources as case 5 showed, but also the perception that the reports are sufficient informative on the state of conservation in order to plan complex interventions.

The cases pointed out a shift in the owner's perception of Monumentenwacht's role. For owners with limited resources to consult specialists, and receiving little support by the municipality or by the Heritage Care Agency, the paradigm has shifted. Monumentenwacht has become in their eyes the main advisor for maintenance and conservation. The good relationship between owners and inspectors, their experience as craftsmen, and insight into the restoration market, strengthens the trust. Moreover, for laymen the difference between signalling and diagnosis is not always clear. The good state of conservation of some cases studied where no conservation specialist was involved showed that this shift is not necessarily negative. Inspectors have knowledge of damage and interventions, and in particular for straightforward causes they are capable of supporting owners to choose a technical appropriate intervention. This can vary from repainting windows, adapting roof details to treatment of wood eating insects. As inspectors are yearly onsite, interventions can be advised timely and therefore preventing the need for more complex interventions. On the other hand, the inspection and reporting methodology of Monumentenwacht was never intended to replace surveys carried out by specialists. For example, Monumentenwacht does not investigate structural problems, and advises owners to consult a structural engineer when related symptoms are noticed. Similarly, Monumentenwacht does not evaluate the effect of deterioration on possible heritage values related to historic matter; this is an evaluation to be carried out by other professionals with insight into the cultural value assessment. However, they do create more awareness among owners about historic details encountered in the building. As Monumentenwacht North Brabant's working methods did not change profoundly since its founding the question arose which factors changed within heritage practice, and whether the cases studies reflect a general trend.

7 CHANGING DUTCH HERITAGE CARE

Heritage care is in several countries a regulated field, with laws addressing listing and conservation. Financial support for owners by tax reduction or subsidies is likewise established in many countries. These boundary conditions change during time, influenced, e.g., by the economic situation of a country, insight into conservation practice and attitudes towards heritage. The role of governmental institutions alters as well, with on one hand a highly controlling and regulating approach, and on the other hand a deregulated heritage care system, with increased responsibilities and powers of owners and believe in a self-regulation of the conservation field.

When Monumentenwacht commenced its works in 1973, the responsibilities of the main key players differed from the present day situation. Restoration architects employed by the Cultural Heritage Agency were involved from the beginning on in the conservation process of (mainly iconic) monuments. This included preliminary meetings where potential restoration architects were discussed, making contra-proposals for interventions, supervising restoration works, and coordinating subsidies. In addition, specialists from the Cultural Heritage Agency were available for more detailed advice. Thus, owners had more governmental support during conservation campaigns. However, in the course of time, the amount of work for the Cultural Heritage Agency increased and the number of staff members was limited. As a reaction, municipalities created their own departments for heritage care. In addition, external restoration architects tended to carry out each time more the tasks originally assigned to the Cultural Heritage Agency, which, according to a witness of this transition (Meischke, 1988), became more bureaucratic and employed less conservation specialists. In 1988, a new heritage law was introduced (Ministerie voor O C en W, 2011), and Dutch heritage care was decentralised, delegating tasks related to permissions and advice to a provincial and municipal level. Inspectors of Monumentenwacht North Brabant working in the field for more than 25 years confirmed a withdrawal of the Cultural Heritage Agency regarding supporting owners on technical problems related to conservation.

Currently, most technical decisions are made without an active involvement of the Cultural Heritage Agency. Some maintenance tasks when using similar materials, detailing and colours can be carried out without requiring a permit. If permission is necessary, a report made by Monumentenwacht can be deemed as sufficient to confirm the technical necessity of an intervention. Understanding changing responsibilities within the conservation process and availability of expertise, explains the shifting role of Monumentenwacht in practice: specialists are considered too expensive and the possible added value of a diagnosis and planning of a treatment is not seen. Awareness is limited when it comes to understanding the deviating aims of technical surveys and the reports based on Monumentenwacht's periodic monitoring, and thus also regarding potential risks.

8 CONCLUSIONS AND DISCUSSION

This multiple case study research on the role of Monumentenwacht in practice showed diverging perceptions of among owners and professionals. Whereas in theory, Monumentenwacht's role is complementary in order to signal threats timely, in Dutch practice this has partially changed. Monumentenwacht North Brabant still fulfils its main aim of supporting owners to carry out maintenance, and its inspection methodology and reporting is in line with this aim. The cases showed that Monumentenwacht can stimulate owners to address maintenance problems timely, which would otherwise have gone unnoticed. However, due to changes within Dutch heritage practice, conservation specialists for diagnosis and treatment are less consulted. Owners tend to refer to the contractors for advice when asking for a quote and thereby avoiding the assumed high costs of advisors. The fact that the state of conservation of most monuments is good indicates that the common shortcut of approaching a contractor directly can lead to acceptable results. Yet more risk awareness is needed, and specialists should be involved when necessary. The observed developments should initiate a wider discussion on the role of Monumentenwacht, and how recent gaps in conservation practice should be addressed: an official extension of Monumentenwacht's tasks and adapting its methodology and organisational setup, a strengthening of the commercial market to lower the threshold for most owners to approach specialists, an increased involvement of controlling or advising institutions, or a combination of the above. The cases indicated that subsidies can stimulate maintenance, yet are not always sufficient to carry out urgent works to prevent decay. Concerns were expressed regarding the subsidy system, as support for restoration works will cease yet degradation can only be delayed.

Most maintenance practice deviates strongly from the ideal process of signalling, diagnosis, treatment, and documentation. Often a sub-cycle of only signalling and treatment is carried out, usually a pragmatic choice and little discussed whether cause and treatment are straightforward, making a timely and cost effective approach justifiable, or whether thereby risks are overseen, either related to an untreated cause, or incompatible intervention. In order to understand whether the changing perception of Monumentenwacht observed in the cases can be generalised and to understand the influence of the different actors active in maintenance, further, quantitative research by means of a survey amongst the members of Monumentenwacht North Brabant has been carried out (Heinemann and Naldini, 2017).

ACKNOWLEDGMENTS

This research was supported by the European JPI Heritage Plus programme and is part of the project CHANGES (Changes in Cultural Heritage Activities: New Goals and Benefits for Economy and Society (2016)). The authors would like to thank Monumentenwacht North Brabant and the owners of the monuments studied for their cooperation.

REFERENCES

CHANGES PROJECT. (2016). *CHANGES (Changes in Cultural Heritage Activities: New Goals and Benefits for Economy and Society) JPI Heritage Plus programme* [Online]. Available: http://www.changes-project.eu/ 2016].

Federatie Monumentenwacht Nederland. (2002). First International Conference Monumentenwacht. 2002 Amsterdam/Amersfoort. Amsterdam: Stichting Nationaal Contact Monumenten.

Heinemann, H.A. & NaldinI, S. (2017). *Vragenlijst CHANGES-Monumentenwacht* [Online]. Available: https://goo.gl/forms/mN7CWtIFbMwXzNlh2.

Meischke, R. (1988). *Beschwouingen over de Nederlandse Monumentenzorg tussen 1918 en ca. 1970,* Amersfoort, Uitgeverij Bekking.

Ministerie Van Onderwijs Cultuur En Wetenschap. (2000). Besluit rijkssubsidiëring grootschalige restauraties BRGR. *BWBR0011525.*

Ministerie Van Onderwijs Cultuur En Wetenschap. (2006). Besluit rijkssubsidiëring instandhouding monumenten BRIM. *BWBR0019439.* The Hague.

Ministerie Van Onderwijs Cultuur En Wetenschap. (2008). Evaluatie Besluit Rijkssubsidiëring Instandhouding Monumenten BRIM.

Ministerie Van Onderwijs Cultuur En Wetenschap. (2016). SIM Subsidieregeling instandhouding monumenten. *BWBR0032075.* The Hague.

Ministerie Voor Onderwijs Cultuur En Wetenschap. (2011). Monumentenwet 1988. In: Justitie, M.V. (ed.). The Hague.

Monumentenwacht Noord-Brabant 2017. Brabantse Monumenten Monitor 2016: Monumentenwacht Noord-Brabant.

Mullink, B. 2010. Restauratie-adviseurs vallen over uitbreiding taken monumentenwacht. *Cobouw,* 10-12-2010.

Naldini, S., Heinemann, H.A. & van Hees, R.P.J. 2017. Monumentenwacht and Preventive Conservation: Changes *In:* Van Balen, K. & Vandesande, A. (eds.) *Reflections on Cultural Heritage Theories and Practices. Vol. 3.* Leiden: CRC Balkema/Taylor & Francis Group.

PRECOM^{3}OS UNESCO Chair. (2016). *What is Preventive Conservation?* [Online]. Raymond Lemaire International Centre for Conservation (KU Leuven). Available: http://precomos.org [Accessed 15-02-2016.

Van Hees, R., Naldini, S. & Nijland, T. (2015). The importance of a Monumentenwacht system—The situation in North-Brabant. Delft: TU Delft, TNO.

Vandesande, A. (2017). *Preventive Conservation Strategy.* Phd, KU Leuven: Faculty of Engineering Science.

Vandesande, A. & Van Balen, K. (2016). An operational preventive conservation system based on the Monumentenwacht mode. *In:* Van Balen, K. & Verstrynge, E. (eds.) *SAHC 2016 Structural Analysis of Historic Constructions. Anamnesis, diagnosis, therapy, controls.* Leiden: CRC Balkema/Taylor & Francis Group.

Vereniging Van Provinciale Monumentenwachten Nederland. (2015). *Inspectiehandboek,* Vught, Vereniging van Provinciale Monumentenwachten Nederland.

Vos, F. (2010). De Monumentenwacht ontstaan, verleden, heden, toekomst. *Vitruvius,* 6.

Innovative Built Heritage Models – Van Balen & Vandesande (Eds)
© 2018 Taylor & Francis Group, London, ISBN 978-1-138-49861-7

Monumentenwacht and preventive conservation: Changes

S. Naldini & H. Heinemann
Faculty of Architecture and the Built Environment, Delft University of Technology, Delft, The Netherlands

R. van Hees
TNO, Netherlands Organisation for Applied Scientific Research, Delft, The Netherlands

ABSTRACT: This paper presents the qualitative research on the importance and function of Monumentenwacht in The Netherlands in relation to the preventive conservation of monuments. The focus was on Monumentenwacht North Brabant. The main part of the research was carried out within the JPI 'CHANGES' project, based on the results obtained in a previous study. The chosen approach was bottom up—monuments, work of inspectors and role of owners—and top down—theory of preventive conservation and role of the Dutch Cultural Heritage Agency (RCE). Interviews showed the appreciation of Monumentenwacht by users of the reports. In the case of non-iconic monuments a shift in the role and activities of Monumentenwacht, from signalling issues to advising owners, was hypothesised, showing an adaptation to changes in Dutch conservation practice. This was confirmed by preliminary quantitative results of an inquiry. Suggestions are made for empowering inspectors and owners by conveying knowledge (from research) to Monumentenwacht.

1 INTRODUCTION

Monumentenwacht rationale conceived in The Netherlands in 1973 is a clear answer to the need for preventive measures ('preventieve maatregelen') (cf. Founding document, article 2, 1973 in Vos, 2010). The process of preventive conservation is referred to in this paper as defined by Vandesande (2017). The process encompasses condition analysis and maintenance in the form of minimal interventions (as little intrusive as possible, meant to avoid major problems), scheduled maintenance and monitoring. The core activities of Monumentenwacht, as carried out for 40 years now, are inherent to the preventive conservation process.

Monumentenwacht was created at a Provincial level as a Public Benefit Organisation (in Dutch ANBI), also having the status of 'Service of general economic interest' (in Dutch DEAB). The main aim of the organisation is to perform inspections on a regular basis (annually or biennially) and promptly signal problems, reported in order of severity. Inspections are visual and cover the whole building, no costly scaffolding is necessary (see also Heinemann and Naldini, 2017a). From the beginning, Monumentenwacht has employed inspectors, who are craftsmen especially trained within the organization in the professions related to the conservation of monuments.

The importance of Monumentenwacht, the 'Monument Watch' organisation for preventive conservation, maintenance and monitoring of heritage buildings in The Netherlands has been the object of a first research (Van Hees et al., 2015b) further broadened and deepened within the JPI project 'CHANGES', which started in 2014 (CHANGES project, 2016). Aim of the 'CHANGES' project is the development and optimisation of preventive conservation, making it widespread and efficient, based on 'best practices' to be distilled from existing conservation practice (Dutch and Flanders Monumentenwacht were studied and compared and the "Distretti Culturali project" (Cultural Districts) in Italy (Fondazione Cariplo, 2016)). It is argued that heritage care should become part of a more profitable and sustainable system, whereby social and financial aspects are considered, like the involvement of general public and stakeholders.

Monumentenwacht North Brabant is an associate partner in the project and at the same time object of the research. The organisation has been given a special status by the Province North Brabant, subsidizing ca. 65% of its income. This set-up allows the organisation to focus on its core business—the inspections and first-aid interventions—and to remain independent. Monumentenwacht North Brabant gives support to the owners, without interfering with architects or advisors. Being an independent organisation, Monumentenwacht's work is highly valued by the government. Based on data obtained by Monumentenwacht's inspections (e.g., Monumentenwacht Noord-Brabant, 2017,

Rijksdienst voor het Cultureel Erfgoed, 2017) the effectiveness of subsidized conservation actions can be assessed and policies can be developed for the monuments in the region.

Also the members feel that their investment in the membership is worthwhile. The balance between costs of the service and quality is considered good, as shown by a satisfaction survey (Vincks et al., 2015). Owners trust inspectors and can easily recur to their support. The membership is generally considered affordable: the yearly membership fee is 55 € and an hourly fee of 28,50 € is charged for inspections and writing reports (Monumentenwacht Noord-Brabant, 2016).

The owners can directly hire a contractor to perform most maintenance activities suggested by Monumentenwacht (without special building permission). This allows prompt execution of essential maintenance actions, limiting their costs and hindering unwanted consequences due to development of damage in the time lapse prior to intervention.

2 METHODOLOGY

The preliminary study (Van Hees et al., 2015b) was used as a basis for the research carried out within the 'CHANGES' project. Theory and practice represented by the propositions of preventive conservation and the role played by the (local) government on one hand and the point of view and experience of Monumentenwacht inspectors in practice on the other, were analysed, being interwoven and affecting one-another. For the definition of preventive conservation it was referred to the PhD thesis of Vandesande (2017). In the first study a qualitative assessment of the importance of Monumentenwacht North Brabant was carried out. The possible consequences of the lack of maintenance were shown. Cases studies and interviews were used to analyse the role and the social impact of the organisation. Within the Changes project, more case studies and interviews were carried out to better understand the actual role of Monumentenwacht in the context of preventive conservation in The Netherlands. A survey was also carried out among owners of the inspected monuments to support the outcomes of the research. Costs of interventions and long term maintenance planning were finally discussed. Possibilities of a closer cooperation between University and Monumentenwacht were explored.

The research on inspected heritage buildings involved confidential matter (reports and actions of owners), which could not be made public. The essence was thus distilled without directly mentioning the source. Except for the interviewed persons in the preliminary study, all the rest are also anonymous. The interviews with heritage professionals and owners were essential to understand the importance and use of the work of Monumentenwacht.

3 DAMAGE SYMPTOMS AND RISK FACTORS

The first study, performed in 2015 (Van Hees et al., 2015b) aimed at showing the importance of maintenance and preventive conservation, carried out thanks to Monumentenwacht, pointing at the problems encountered by the inspectors and showing the societal and practical role of Monumentenwacht for people involved in the conservation of monuments. Special attention was given to the inspectors and professionals working in the field of monument conservation. Prior to the analysis, inspectors were asked to make a list of the failures they found in their daily practice and divide them in four categories (in relation to frequency,

Figure 1. Examples of the lack of maintenance and moisture ingress. Left: Lack of maintenance wood (gutters) and missing slates. Middle: Rotting wood of a window frame. Right: Rotting wood and white rot (dry rot fungus) on a timber ceiling (pictures from Van Hees et al., 2015b).

danger, importance, complexity of solution). The relevant problems encountered were further grouped in terms of urgency that is to say from safety issues to minor defects. The safety issues clearly need immediate intervention or solution and no discussion was deemed necessary. The shortcomings (failures) of materials listed, instead, were further studied in terms of future development and potential risk. With the help of the FMEA (Failure Mode and Effects Analysis) method, used in a qualitative way, the consequential damage of an initially small problem was made clear; as well as the much greater effort needed to repair the consequential damage at a later stage.

The examples in Figure 1 show the escalation of damage severity: problems immediately solvable with ordinary maintenance (failure of roof slates and gutters), can lead, if neglected, to ingress of water and damage; in the most severe stage shown, wood is attacked by dry rot fungus and needs removal and measures against ingress of moisture are necessary. It is evident that in the conservation of monuments different factors will determine the importance of maintenance and the economic and material loss in case of postponed or neglected intervention (e.g. moisture ingress in a plain masonry wall vs masonry with stucco or fresco decoration).

4 SUPPORT AND CONTROL FUNCTIONS

The most important characteristics of Monumentenwacht were identified in the prompt signalling of damage or failure, in the first place, the first-aid repairs limiting or eliminating damage, the support given to the owners in understanding the necessary interventions, the control of the work done by contractors and the monitoring whereby the effectiveness of the interventions after 1 or 2 years' time is assessed. The maximum time span of 2 years is considered essential to optimize monitoring and to control inspections. Control of the quality of the therapy is considered problematic in the analysis of Monumentenwacht Flanders (Vandesande and van Balen, 2016). Maybe one cause is the lapse of time between inspections, 4–5 years. In North Brabant the controls occurs immediately, and if a contractor has performed poorly or has made divergent choices from what was planned, correction is required. The trust between inspectors, owners, and contractors, and their frequent co-operation lies at the basis of a good functioning system.

About the first-aid repairs, the case illustrated in Figure 2 shows the importance of a correct identification of the cause of the observed damage and a prompt and suitable intervention. The detailing of a lead sheet in a roof was responsible for leakage (another visible damage symptom was the growth of plants in the mortar of the masonry wall). The location was hardly accessible and the problem was easily solved when the inspectors reached the place: they simply modified the shape of the sheet and hindered the flow of the water along the wall whereby also the biological growth ceased (Van Hees et al., 2015a).

5 REVIEW OF MONUMENTNEWACHT'S REPORTING SYSTEM

The results of the inspections are presented in the form of a Monumentenwacht report, and the six condition scores are based on the standard for condition assessment NEN 2767:2013. The findings are highlighted with colours for immediate understanding of their severity grade and prioritized within a long-term plan. The reports evolved during the last 40 years, and are now clearly written and furnished with more pictures then in the past, thus easy to understand for the owners, who,

Figure 2. Effect of a minor intervention carried out by Monumentenwacht North Brabant: left situation as found with biological growth on the wall; middle: intervention carried out—folding of the lead sheet to deviate the water flow; right: situation afterwards showing the effectiveness as no biological growth reappeared (pictures from Van Hees et al., 2015b).

in case of doubt or questions can ask the inspectors for informal explanations. In the new type of report some warnings are added, dictated by the long experience of the inspectors in the field and the many wrong interventions seen. The structure of the reports is thus clear. The content was further examined. The reports are based on the visual assessment of the state of preservation/damage of the buildings and thus a diagnosis of the causes of damage is not available, unless the damaging mechanisms are evident. A number of reports were studied and three case studies were carried out—involving Master's students of the Faculty of Architecture and the Built Environment, TU Delft (course 'Building Conservation Assessment'), to better understand the level/completeness of the inspections of Monumentenwacht confronting their findings with the situation found in practice. Monumentenwacht signals damage and rates it in terms of urgency of intervention—purely based on visual inspection (Van Hees et al., 2015b).

So far, the Monumentenwacht inspectors can be compared with the general practitioner in health care in The Netherlands, who, having examined the patient (symptoms of disease), prescribes a treatment (or perhaps no treatment) to solve clear cases or suggests a specialist recourse in complex cases, to come to a sound diagnosis and a therapy. If the patient decides to call on a specialist, the Dutch health insurance system will provide economic support. In the case of heritage care, private owners of a dwelling can only deduce their expenses for maintenance from their income tax. No facilitation is available for inspections.

The therapy is represented by the prioritised interventions suggested, whereby the inspectors support the owners to evaluate the specifications of the contractors' plans. Controls of the works are always done in the monitoring phase and often immediately after an intervention. Differently from other countries, in The Netherlands many maintenance activities can be carried out without obtaining permission from the Dutch Cultural Heritage Agency or the municipality. This allows prompt execution of essential maintenance actions, limiting the costs and hindering unwanted consequences due to development of damage in the time lapse prior to intervention. The controls of the contractor's work is also a means for limiting costs and loss of historic material.

Within the 'CHANGES' 16 projects were selected and further studied. The criteria and the conclusions of the case-studies have been explained elsewhere (Heinemann and Naldini, 2017a). A clear difference in terms of approach to conservation was observed between privately owned dwellings or other non-iconic buildings and iconic, well-known buildings.

The responsibility of the interventions in The Netherlands is not necessarily entrusted to an architect, like in other European countries as Italy, or to a supervisor of the Dutch Cultural Heritage Agency (RCE). Due to decentralisation of tasks and responsibilities, in the case of non-iconic buildings, the RCE neither performs controls on plans and interventions nor bears any responsibility. The municipality under which the heritage buildings fall, does often not count sufficient experts in monuments among its employees.

Under circumstances Monumentenwacht is trying to offer more support to the owners and avoid failures. Recalling the comparison with general practitioners, the creation of a Monument Health system could change the way of operating of the various actors and include assistance of experts

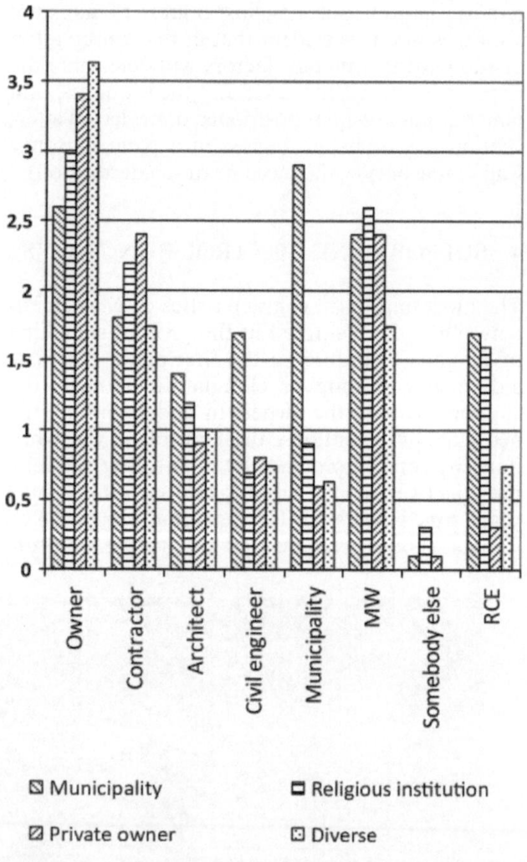

Figure 3. Degree of influence different parties, shown in the x-axis, have on the choice of an intervention based on an enquiry among members of Monumentenwacht North Brabant (n = 192); y-axis: 0 = not involved; 4 = strong influence. The responds are divided into the type of ownership: municipalities, religious institutions, private owners and diverse forms of ownership.

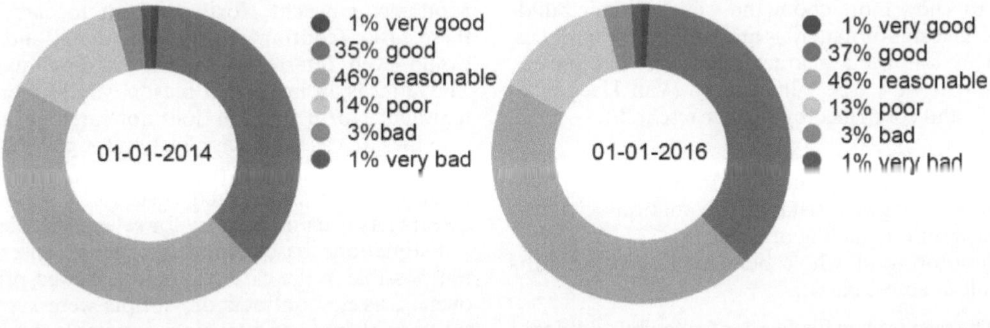

Figure 4. Overview of the state of conservation of objects inspected by MW North Brabant in 2014 (left) and 2016 (right) (based on Monumentenwacht Noord-Brabant, 2017).

and financial support for laboratory investigations. The inspectors should make clear to the owners that—in complex cases—specialized advice and in depth material investigations result in sparing money and historic materials.

To verify the hypothesis of the changing role of Monumentenwacht within a changing system, an enquiry was carried out among the members of the organisation, that is to say the owners of the monitored buildings (Heinemann and Naldini, 2017b). The preliminary quantitative results of the inquiry confirm the hypothesis, and that a.o. Monumentenwacht has a strong influence on owners' decisions (Figure 3).

In how far the owners of non-iconic buildings and private dwellings follow the advice of Monumentenwacht strongly depends on their financial means. Normally though, the urgent interventions are done in time. The buildings inspected by Monumentenwacht (3000 per year) remain in a relatively good state of preservation (Monumentenwacht Noord-Brabant, 2017), which is actually the aim of planned maintenance and monitoring (Figure 4).

6 INTERVIEWS

Interviews were done to gain better insight into how the monument world perceives Monumentenwacht. Asked for an interview were: a representative of the foundation ERM (Foundation 'Erkende Restauratiekwaliteit Monumentenzorg', acknowledged restoration quality in monument care), a restoration architect, the directors of an insurance company, specialized in insuring monuments (including most churches in The Netherlands), an important restoration contractor, a representative of the Cultural Heritage Agency and a private monument owner (Van Hees et al., 2015b). The interviews were meant to be explorative and not exhaustive, due to the restricted number of the interviewed persons. However, these were all professionals (with the exception of the owner), all exploiting important activities in relation to monuments.

'10 questions were formulated to serve as a basis for the conversations, namely:

1. *How do you know (of) MW?*
2. *In your opinion, in which kind of activities is MW engaged?*
3. *Which of the activities do you consider: necessary-useful-unnecessary*
4. *Do MW's activities overlap with those of other organisations you might know of?*
5. *What is the relationship architects—MW?*
6. *In your opinion, what should/could the relationship of MW and (standardized) quality systems be?*
7. *In your opinion, what is the added value that MW offers?*
8. *Do the repeated inspections by MW (monitoring) offer added value?*
9. *How could MW's quality be improved?*
10. *Which other tasks could/should MW carry out?'* (Van Hees et al., 2015b).

The answers given pointed at the following: architects find Monumentenwacht baseline measurements important when starting a project. The Province considers inspection reports essential for policy making and subsidizing heritage conservation. Contractors find the clear and well-defined role of Monumentenwacht a means of work creation, as they are hired for the necessary interventions. For them the report can be a basis for assessing the amount of work to be done and its costs. For the insurance company, dealing with monuments and most churches in the country, the reports of Monumentenwacht are welcome, as its own inspectors do not generally have the possibility of assessing the state of roofs and other less accessible parts of a building. Owners require the reports for maintaining their properties: they not only understand what needs to be done, but also get to know more about the value of their buildings, in terms of materials and building techniques.

The following characteristics of Monumentenwacht were especially praised (Van Hees et al, 2015) and confirmed by further research:

- independent organisation
- reliable and experienced craftsmen
- inspectors give assistance/explanations
- clear aim: signalling of damage
- monitoring of whole building (including difficult to access parts).

The results were further confirmed by additional interviews carried out during the 'CHANGES' project. These were done on the same basis to representatives of Dutch conservation world. The interviews helped understanding the role and potentials of Monumentenwacht and figure out possible developments.

7 COSTS AND LONG TERM MAINTENANCE PLANNING

Monumentenwacht North Brabant does not calculate the costs of maintenance for the owners. However, from 40 years' experience, Monumentenwacht has gained empiric knowledge on the yearly budget an owner should consider depending on the state of conservation of the building. For maintenance of a house or farm which is in a reasonable to good state, an owner should expect to spend on average 1–1, 25% of the market value of the house (which is based on the council tax) for a period of 10 years. In case of overdue maintenance, resulting in a mediocre/poor state of conservation, one should expect costs 3–5% of the market value per year as yearly maintenance costs. In case the building is in a poor condition, the costs can be 25% of the market value, and not only maintenance but restoration could be necessary (Vereniging van Provinciale Monumentenwachten Nederland, 2015). Exact calculation of costs and thus evaluation of the performed preventive conservation in economic terms, is not possible (cf. also Vandesande, 2017). Although quantitative assessment of costs of the 'avoided restoration' thanks to preventive conservation is not possible, the Monumentenwacht rule of thumb in combination with the possible development of damage, when performing the FMEA analysis, gives a qualitative, but very clear indication of the positive effects of preventive conservation.

In The Netherlands, the long-term maintenance planning and the calculation of costs are established fields within the conservation practice (and building maintenance in general). These tasks are carried out by architects and consultants for maintenance. Therefore, there is less need for Monumentenwacht North Brabant to carry out these tasks (contrary to e.g. Flanders), and this would even interfere with the existing market. The latter is an important reason why Monumentenwacht North Brabant does not carry out these activities. In case of an application for governmental maintenance subsidies, a planning of maintenance works and related costs for a period of 6 years has been mandatory for several decades.

Estimations based on costs and expenses are not possible in the case of the interviewed private owners, as no contracts or receipts were kept. In the case of large and iconic monuments there are documents of the expenses, but it is not clear what costs have been made for prevention and what for restoration. Furthermore, in retrospective it is not possible to determine whether the interventions were technically correct, or if circumstances led to the choice of a more expensive or technically inferior solution (e.g., emergency repair, availability of contractors, additional interventions to improve the building).

To draw a relationship between expenses and state of conservation, an exact description of the interventions in terms of aims (after a diagnosis), materials used, application techniques and execution should be available. Only then the cost—effectiveness of an intervention can be assessed.

8 SCIENCE AND PRACTICE BRIDGED

Within the 'CHANGES' project a close cooperation has developed between Monumentenwacht and TU Delft, aiming at the dissemination of knowledge on topics that the inspectors and owners deem relevant. Based on the issues found in practice, the results of research projects in the field of conservation can be made available and disseminated to the inspectors, who can increase the awareness of the owners. Also modern materials, like concrete, and their failures could be included. Further, the use of a standardized terminology for the identified damage symptoms in practice could be the basis for a diagnosis and would improve comparisons and monitoring. TU Delft has suggested the terminology of the Monument Damage Conservation System (MDCS, TNO et al., 2017): each type of damage and symptom are related to one term and definition only, and possible damage mechanisms, and can be identified on the mere basis of a visual inspection.

The empowerment of inspectors and owners is fundamental within the Dutch context, where the decision on most actions is entrusted to them. The long lasting relationship with Monumentenwacht offers the possibility to involve owners more actively in the management and maintenance of their buildings. One important point is the archiving of documents related to inspections and interventions.

Most case studies showed that no 'patient's record' of the building exists, which would be essential for monitoring and planning maintenance and also for cost related studies: it would be advisable to ask the owners to upload all documents concerning the work done by contractors on e.g. a Monumentenwacht platform. Monumentenwacht in the Province of Gelderland is developing an app for contractors to keep a list of specifications and costs.

The structural involvement of Monumentenwacht in the re-use of abandoned young monuments, like industrial buildings, could mean preventing further damage until the building is restored, thereby saving costs, and including preventive maintenance aspects in the design phase. Monumentenwacht could also identify dangerous materials, and limitations for old materials and techniques meant to be kept or re-used in the new design.

The monitoring task of Monumentenwacht should be made standard, and could include, besides the usual activities, the monitoring of all measures taken to upgrade the building to suit new requirements (Wilkinson et al., 2014), improve its energy consumption and make it more suitable for modern life.

9 CONCLUSIONS

Research has been carried out on the importance and role of Monumentenwacht in Dutch heritage conservation. The first, qualitative and quantitative results will contribute to achieve the aims of the 'CHANGES' project. Quantitative assessment of costs of the 'avoided restoration' thanks to preventive conservation is not possible. However Monumentenwacht rule of thumb in combination with the possible development of damage, when performing the FMEA analysis, gives a qualitative, but very clear indication of the positive effects of preventive conservation.

The limited costs of membership and activities make it possible for Monumentenwacht to guarantee the regular monitoring and to maintain or improve the state of conservation of a high number of monuments. Monumentenwacht North Brabant does not provide private owners (not iconic buildings) with budget planning for maintenance, which is mostly entrusted to the contractor. When applying for financial support the owners may recur to an architect. The work of Monumentenwacht has proved to be profitable for the general level of conservation of the heritage buildings in The Netherlands, and consequently for the image of the country. A shift in the original role of the inspectors was observed, towards the fulfilment of an advisory task. Empowerment of inspectors and owners, disseminating knowledge from university research was discussed and found feasible and productive. Next to their technical tasks the inspectors play a societal important role making owners aware of the material and technical value of their buildings and involving them in conservation. Monumentenwacht has many potentialities, especially thanks to the good structure of the organization and the consensual appreciation. It might be profitable and beneficial to make Monumentenwacht more than just an optional link in the conservation cycle of monuments, and also involve the organisation in the adaptive re-use of heritage buildings.

ACKNOWLEDGEMENTS

This research was partially supported by the European JPI Heritage Plus programme and is part of the project CHANGES (Changes in Cultural Heritage Activities: New Goals and Benefits for Economy and Society). The authors would like to thank Monumentenwacht North Brabant, owners and interviewees for their cooperation.

REFERENCES

CHANGES PROJECT. (2016). *CHANGES (Changes in Cultural Heritage Activities: New Goals and Benefits for Economy and Society) JPI Heritage Plus programme* [Online]. Available: http://www.changes-project.eu/2016.

Fondazione Cariplo. (2016). *Cultural Districts* [Online]. Available: http://www.fondazionecariplo.it/en/projects/arts-and-culture/distretti-culturali/distretti-culturali.html [Accessed 24-11-2016 2016].

Heinemann, H.A. & Naldini, S. (2017a). The role of Monumentenwacht: 40 years theory and practice in The Netherlands. In: Van Balen, K. & Vandesande, A. (eds.) *Reflections on Cultural Heritage Theories and Practices. Vol. 3.* Leuven: CRC/Balkema.

Heinemann, H.A. & Naldini, S. (2017b). *Vragenlijst CHANGES-Monumentenwacht* [Online]. Available: https://goo.gl/forms/mN7CWtlFbMwXzNlh2.

Monumentenwacht Noord-Brabant (2017). Brabantse Monumenten Monitor 2016 Vught: Monumentenwacht Noord-Brabant.

Monumentenwacht Noord-Brabant. (2016). *Monumentenwacht Noord-Brabant* [Online]. Available: http://www.mwnb.nl [Accessed 11-04-2016].

Nederlands Normalisatie-Instituut (2013). NEN 2767-1+C1 Condition assessment—Part 1: Methodology. Delft: Nederlands Normalisatie-instituut.

Rijksdienst Voor Het Cultureel Erfgoed. (2017). *The Heritage Monitor* [Online]. Amersfoort. Available: https://erfgoedmonitor.nl/databank [Accessed 04-07-2017 2017].

TNO, Rijksdienst Voor Het Cultureel Erfgoed & TU Delft. (2017). *MDCS/Monument Diagnosis and Conservation System* [Online]. Available: http://mdcs.monumentenkennis.nl/ [Accessed 24-10-2017].

Van Hees, R., Naldini, S. & Nijland, T. (2015a). Het belang van de Monumentenwacht—De situatie in Noord-Brabant. Delft: TU Delft, TNO.

Van Hees, R., Naldini, S. & Nijland, T. (2015b). The importance of a Monumentenwacht system—The situation in North-Brabant. Delft: TU Delft, TNO.

Vandesande, A. & Van Balen, K. (2016). An operational preventive conservation system based on the Monumentenwacht model. In: Van Balen, K. & Verstrynge, E. (eds.) *Structural Analysis of Historical Constructions—Anamnesis, diagnosis, therapy, controls.* Leiden: CRC Balkema/Taylor & Francis Group.

Vandesande, A. (2017). *Preventive Conservation Strategy.* PhD, KU Leuven: Faculty of Engineering Science.

Vereniging Van Provinciale Monumentenwachten Nederland 2015. *Inspectiehandboek,* Vught, Vereniging van Provinciale Monumentenwachten Nederland.

Vincks, K., van Dalen, B. & van Mol, M. (2015). Klanttevredenheid Monumentenwacht Noord-Brabant Klanttevredensheidsonderzoek 2015. Tilburg het PON.

Vos, F. 2010. De Monumentenwacht ontstaan, verleden, heden, toekomst. *Vitruvius,* 6.

Wilkinson, S., Remøy, H.T. & Langston, C. (2014). Sustainable building adaptation: innovations in decision-making. Chichester, England: Wiley-Blackwell.

Innovative Built Heritage Models – Van Balen & Vandesande (Eds)
© 2018 Taylor & Francis Group, London, ISBN 978-1-138-49861-7

Implementation of maintenance systems: Identification of maintenance practices within the Monumentenwacht model in Belgium

N. Van Roy & E. Verstrynge
Department of Civil Engineering, KU Leuven, Heverlee, Belgium

A. Vandesande & K. Van Balen
Raymond Lemaire International Centre for Conservation and Department of Civil Engineering, KU Leuven, Heverlee, Belgium

ABSTRACT: As part of the CHANGES-project, a research project within the European Joint Programming Initiative on Cultural Heritage, maintenance practices are studied in Belgium for the urban region of Mechelen based on the study of seven cases. The main goal of the research, which is performed in collaboration with Monumentenwacht Flanders, is to identify maintenance practices and to understand the types of interventions that are executed. For all the seven selected cases discussed in this paper, the owner is a member of Monumentenwacht Flanders. Maintenance practices were detected based on social research methods, a technical analysis of the performed interventions, and a statistical analysis of the inspection reports of Monumentenwacht. This mixed method analysis revealed that owners take the severity of damages that are detected by Monumentenwacht into account when they decide on maintenance actions, but that their actions are also influenced by practical considerations.

1 INTRODUCTION

Within the CHANGES (Changes in Cultural Heritage Activities: New Goals and Benefits for Economy and Society) project existing models within the framework of Planned Preventive Conservation are analysed with the aim of demonstrating how such models contribute to local sustainable development (CHANGES 2017). For the Belgian context, the analysed model is Monumentenwacht Flanders. Monumentenwacht is an organisation in Flanders that assists owners with regular condition assessments and technical advice for the maintenance of their historic buildings. They furthermore generate awareness on the importance of maintenance for historic buildings. Each of the five provincial Monumentenwacht associations are integrated into the services of their provinces in Flanders, while one umbrella organisation oversees uniformity and the quality of the provided services. The so called Monumentenwachters carry out inspections and generate a report that contains a detailed condition assessment and recommendations for the actions that the owners should undertake with an indication of their urgence and priority (Vandesande 2017). The Monumentenwacht model therefore entails an approach where experts regularly inspect buildings, detect damages and inform owners on the necessary actions to be taken. Within this framework, the owners decide on the frequency of inspections and their response to detected damages and recommendations. This research therefore investigates how private owners operate within the Monumentenwacht model.

The investigation of current practice is done based on case studies, which were selected in a urban and a rural area. The results in this paper focus on the urban area of Mechelen, where eleven cases were selected. All the buildings are listed heritage and all the owners are member of Monumentenwacht Flanders. Seven of the eleven cases are owned by private owners, while the other 4 cases have public ownership. This investigation focuses on the maintenance practices of the private owners.

2 METHODOLOGY

In order to investigate existing maintenance practices in Flanders, a methodology was developed that consisted of in-depth interviews with seven private owners, a quantitative data analysis of the reports of Monumentenwacht for four of the seven cases, and a time-series analysis of two of those four case studies. The investigation was founded on a state of the art on maintenance approaches for built heritage. The state of the art departs from the added value of preventive conservation practices and identifies how maintenance practice should ideally be conducted within this framework.

The in-depth interviews with seven private owners allowed to identify the owners' maintenance objectives and approaches. The obtained data was analysed based on Grounded Theory, a research methodology of social sciences. The method starts from the idea that the researcher develops theory based on empirical material. Furthermore, Grounded Theory expresses the importance of a cyclic work approach in the analysis of data. This analysis strategy enables open data retrieval, and a gradual building up of a validated hypothesis. After interviewing the seven private owners, transcribed interviews were therefore analysed through a three step coding method. A first step is open coding, an analytical process to identify concepts. A second step is axial coding, a process of linking the identified concepts into categories and subcategories. A third step is selective coding, wherein the concepts are integrated into a final hypothesis (Mortelmans 2013, Glaser & Strauss 1967).

The added value of this research is that the owners' objectives and approaches are supplemented with a more detailed research of the actions of owners that are part of the Monumentenwacht model. Four of the seven private owners have regular inspections done by Monumentenwacht and can thus be considered active participants in the model. The research focused on the response of the owners to the inspection reports based on a quantitative analysis of the inspection data. A bivariate analysis investigated the relation between the response time and the severity of damages. The scatter in the data demonstrated that the relationship was influenced by other parameters. These influencing parameters were identified and validated based on a time-series analysis of two of the four case studies. The applied time-series analysis consisted of the compiling of chronological events (actions, inspection data, interventions) as a means to investigate causalities and the response of owners to the inspection reports (Yin 2014). The analysis was applied for two cases that had a relevant amount of project documents available concerning the interventions that were executed based on the inspection reports.

A hypothesis on the maintenance practices of private owners in Flanders was drafted based on the combined results of qualitative social research and quantitative data analysis of technical documents.

3 MAINTENANCE APPROACHES FOR BUILT HERITAGE: STATE OF THE ART

Within the framework of preventive conservation approaches, literature emphasises that maintenance approaches for built heritage should entail a combination of monitoring, planning and technical actions, that are part of a well-defined plan (Forster & Kayan 2009, Dann, Worthing, & Bond 1999, Vandesande 2017). The main aim is the preservation of cultural significance (ICOMOS 1999, Feilden & Jokilehto 1998, Van Balen 2015, Dann 2013, Forster & Kayan 2009). Other aims of maintenance activities for historic buildings that were identified in the literature are the preservation of historic fabric (Van Balen 2015, Dann 2013, Della Torre 2013, Forster & Kayan 2009, Dann & Cantell 2005, Dann &Wood 2004), the preservation of the building's use or function (Dann 2013, Dann & Wood 2004), long term cost saving (Van Balen 2015, Dann 2013, Forster & Kayan 2009, Dann & Cantell 2005), minimization of uncertainties (Elena Charola, Henriques Fernando, & Magadan Marcelo 2015, Forster & Kayan 2009), community involvement (Van Balen 2015, Della Torre 2013, Van Balen & Vandesande 2013), capacity building (Moioli 2013, Della Torre 2013) and social benefits related to local sustainable development (Van Balen & Vandesande 2013, Forster & Kayan 2009).

All interventions should be minimal technical interventions, that follow a four step approach: anamnesis, diagnosis, therapy, controls (ICOMOS 2003, Vandesande 2017). The first step, the anamnesis, aims at gathering information on the building, its components, its materials, and its environment. The aim of the diagnosis is to understand detected damages and identify their causes (Van Balen 1988). The therapy should be based on the diagnosis and should aim at threating the causes, rather than the symptoms. Interventions should respect conservation principles, which implies that interventions should respect the principle of minimal intervention doing as little as possible and as much as necessary as well as the principles of reversibility or compatibility and retreatability (ICOMOS 2003). The aim of the controls is to evaluate the interventions, to perform a quality check during execution, to predict the future behaviour and to check the behaviour through long term monitoring (Van Balen 1988). Therefore, interventions should be accompanied by controls during the works, and monitoring should be carried out after the works to ascertain their efficacy (Verstrynge, Schueremans, & Smars 2012). Monitoring activities before, during and after the works should be documented (ICOMOS 2003). In the Flanders context, the anamnesis and the controls are integrated in the Monumentenwacht model.

Vandesande refers to the importance of knowledge within preventive conservation practice. This includes knowledge on performances, materials, techniques and damage processes, as well as know-how on the application of interventions in practice and workmanship (Vandesande 2017). For the same reasons, Marconi referred to the art and culture of maintenance of historic buildings in his works during the 1980s. He describes a practice of maintenance activities that tends to get lost in the final epochs of the 20th Century due to the belief in miracle products and solutions that can protect the building's outer shell from the influences of its environment (Marconi 1991).

It can be concluded that for built heritage, maintenance activities should be integrated in a well-defined approach and aim at the preservation of cultural significance. The applied approach should entail regular monitoring and small interventions that follow a four step approach: anamnesis, diagnosis, therapy, controls. Knowledge exchange, documentation, knowledge enhancement and capacity building are essential aspects within the framework of maintenance activities on historic buildings.

4 IDENTIFICATION OF THE MAINTENANCE OWNERS' OBJECTIVES AND APPROACHES BASED ON IN-DEPTH INTERVIEWS

Seven of the 40 private owners that are member of Monumentenwacht in Mechelen were selected for the identification of the owners' objectives and approaches. A statistical analysis of the inspection reports was done for four of the seven cases (cases 2,5–7). Finally, a time-series analysis was done of two cases, (cases 2 and 5).

4.1 *Identification of the maintenance owners' objectives*

In order to understand the aim of the owners' maintenance approaches, owners were questioned during in-depth interviews concerning their appreciation of the property, as well as their perception on maintenance. Concerning the owners' appreciation of their properties, three maintenance attributes were detected. Firstly, there is a distinct difference between a perception that focuses on the concepts of the architectural values and a perception that takes into account the importance of the authenticity of historical fabric. The only owners who explicitly referred to the authenticity of historical fabric are those of cases 2, 3 and 5. Secondly, most of the owners demonstrate a strong emotional bond with the property. For case 1, 2, 3, 4 and 7 the strong emotional bond was present from the beginning, while for case 5 this bond grew over the years. The owners of case 6 demonstrate less emotional involvement, since they bought the property as a financial investment. Thirdly, most of the owners had to invest a large initial sum for the restoration of the properties, that were often in bad state of preservation. Only the owners of case 2 indicated that the property was in a good state of preservation and they did not undergo a restoration process. The restoration process was a long and intense experience for most of the owners, which left them with a conviction to avoid future deterioration due to neglect.

Concerning the owner's perception on maintenance, six of the seven owners indicated that they do not consider the maintenance of their buildings to be specifically different than that of other buildings (cases 1–4,6–7). They all consider the goal of maintenance to be the preservation of the building's state. Four owners (cases 1–3,6) approach this preservation from a preventive viewpoint, and they indicate that maintenance should avoid the occurrence of damage, based on regular inspections and small maintenance works. For two of them (cases 2,3), the goal is also to maintain the historical fabric. Three owners (cases 4,5,7) do not solely perceive maintenance as the preventive actions to avoid damage, but they also include repair works, restoration works and renovations or improvements in their perception on maintenance.

Based on these insights, six maintenance attitudes were defined: (1) concern for historical fabric, (2) emotional bond, (3) prior investment, (4) property in bad state of preservation when purchased, (5) maintenance is evident, (6) preventive approach. The maintenance objectives are summarized for each case in Table 1. It is important to note that for the selected case studies, the financial burden of regular works was not identified by any of the owners as impeding them to regularly undertake maintenance activities.

4.2 *Identification of the maintenance owners' approaches*

Six of the seven owners have a well-defined approach for the maintenance of their property. Four of them (cases 2,5–7) already rely on the

Table 1. The maintenance objectives.

	C1	C2	C3	C4	C5	C6	C7
Concern for historical fabric		x	x				
Emotional bond	x	x	x	x	x		x
Prior investment	x		x	x	x	x	x
Bad state of purchased property	x		x		x	x	x
Maintenance is evident	x	x	x	x		x	x
Preventive approach	x	x	x				

Monumentenwacht model, which means that they contact Monumentenwacht for regular inspections and contact a contractor after selecting a number of recommendations they want to act upon. The average frequency (f) of inspections is once every 3.9 years (case 2: average time interval is 5.33 years, case 5: average time interval is 3.5 years, case 6: average time interval is 3.8 years, case 7: average time interval is 3 years). The owner of case 1 purchased the property three years ago and is still in the process of restoring it, but plans to apply the same approach based on a two to three yearly inspection of Monumentenwacht. The owner of case 1 is the only one who also plans to involve an architect in the maintenance interventions. The owner of case 3 does not rely on regular inspections by Monumentenwacht, but performs the inspections on his own. A maintenance calendar for the house was developed over the years for regular actions and small interventions that include check-ups and cleaning, while larger interventions are done during a two weeks maintenance period in the summer, such as painting of timber joinery. The owner of case 4 has no well-defined approach and relies on the durability of previous repairs and renovation works.

Three of the seven owners (cases 2–4) indicated they had difficulties in finding a good contractor to execute works. Two owners (cases 1,5) found contractors through a local architect with experience in conservation of historic buildings. Five owners (cases 2,3,5–7) indicate that they mostly find contractors through their personal network.

On the one hand, it is difficult to find contractors who are specialized in specific works, as indicated be the owners of cases 2–4. The owner of case 2 also indicates that local contractors are often not specialized and distant contractors do not want to make the journey for a small repair. Furthermore, specialized contractors are often occupied on large projects and do not accept small projects, as indicated by the owners of cases 2 and 4.

On the other hand, it is difficult for owners to understand when a contractor is up for the job. It is not always clear when the owners should contact specialized craftsmen. The owner of case 3 also indicated that the costs of specialized works are high. Owners would like to get more support in understanding how to do the works, and they would like to get this information from someone who is objective. There is furthermore a lack of someone who can support them in the technical aspects and who can guide them in the complexity of the collaboration and coordination of different stakeholders. Finally, due to bad experiences (lack of motivation and bad intents), five of the seven owners have little confidence in the construction practitioners, which includes both architects and contractors.

5 IDENTIFICATION OF TENDENCIES IN THE OWNERS' MAINTENANCE APPROACHES BASED ON A STATISTICAL ANALYSIS OF INSPECTION REPORTS

Four of the seven owners (cases 2,5–7) acquired regular inspection reports of their property during a period of time in which they also performed maintenance interventions. Table 3 gives an overview of the respective time frame, the inspection dates, the number of inspections, the average time interval and the average frequency for the four selected cases.

The response of the owners to the reports is investigated with a focus on the rainwater disposal system, consisting of roof coverings, elements for the rainwater disposal (gutters, drainpipes), and the connections between elements. Two aspects were investigated: the owners' response time in relation to the recommendations of Monumentenwacht, and the owners' response time in relation to the severity of the detected damages.

A bi-variate analysis of the relation between the urgency (based on recommendations Monumentenwacht) and the owner's response time, shows a slight tendency towards faster response time for more urgent recommendations, as seen in Figure 1. The urgency of the recommendations is based on four scores: (1) no recommendations made, (2) recommended to solve on the long run, (3) recommended to solve shortly, (4) recommended to solve urgently. The owner's response time is determined based on the inspection reports. This means that the time interval is calculated as the time difference in years between the moment when the damages were first detected and the moment when they were no longer appearing in the reports. This means that the calculated response time will be influenced by the accuracy of the inspections and their frequency.

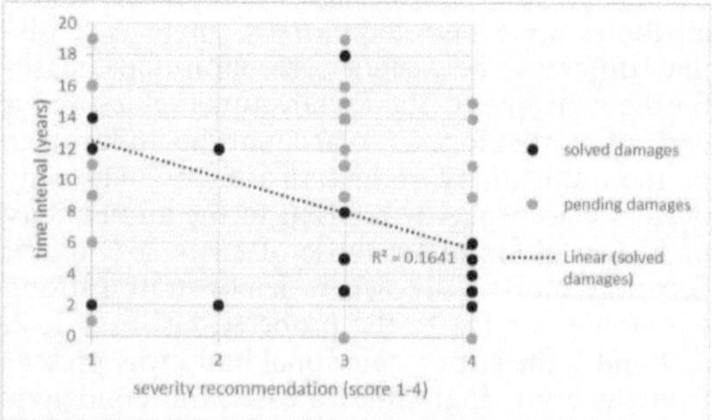

Figure 1. Relation between the urgency of the recommendations and the owner's response time, based on data Monumentenwacht Vlaanderen (MOWAv).

The severity of the damages was defined based on the type of damage, the condition and the probability of resulting infiltration (inf.). All damages were classified according to four categories: (1) disintegration of materials and connections, (2) poor design or execution, (3) mechanical damage to materials or loosening of connections, (4) missing elements. Based on the reports of Monumentenwacht, the probability of resulting infiltration was quantified for the four damage categories. The probability P (inf.) was calculated as the number of occurring damages (NI) leading to infiltration divided by the total number of occurring damages of a damage type (TO). The total number of occurring damages was recorded by summing up all the occurrences in the reports of Monumentenwacht for the seven selected cases. This calculation allowed to classify and score them according to their severity (Table 2). The severity of an occurring damage is then defined as the sum of the damage category's score and the damage condition's score. The latter was based on the assessment by Monumentenwacht and was thus found in the inspection reports. A relation was found between the probability of resulting infiltration and the condition score, see Table 2.

A bi-variate analysis of the relation between the severity, as defined based on the type of damage and the condition and the owner's response time, demonstrates again a slight tendency for solved damages to faster solutions for more severe damages, as can be seen in Figure 2.

The scatter in the results of both bi-variate analyses indicates that the relationship between the severity and the owners response time is moderated by other factors. In order to understand these factors, the types of interventions that were resolved within a shorter time frame were analysed. Table 4 gives an overview of the damages that were

Table 2. The probability of resulting infiltration P was calculated by dividing the Number of Infiltration (NI) by the Total Occurrences (TO) of a damage category and a condition score, based on data MOWAv.

Damage category	NI	TO	P
Disintegration	3	17	8.33%
Poor design or execution	1	12	17.65%
Mechanical damage, loosening	11	24	45.83%
Missing elements	3	5	60.00%
Condition	**NI**	**TO**	**P**
Good	0	0	0.00%
Tolerable	2	31	6.45%
Moderate	9	19	47.37%
Bad	7	8	87.50%

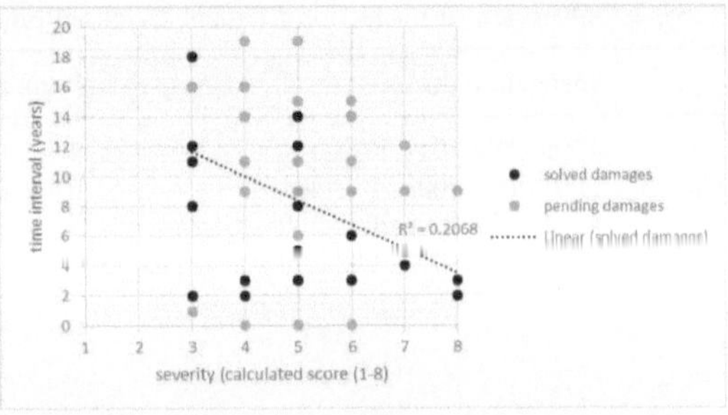

Figure 2. Relation between the calculated severity and the owners' response time, based on data MOWAv.

solved within 5 years time. At first, the influence of the accessibility was investigated. The last column of the table indicates the accessibility of the location of damage. Difficult accessibility means that scaffolding is needed to access the location for works. Medium accessibility means that a ladder is sufficient for accessibility. Easy accessibility indicates that the location can be accessed without ladders or scaffolding. Five of the thirteen solved damages were found in easy accessible areas. The six damages that were in difficult accessible areas were solved as part of a larger repair intervention. For case 5, this entailed the renewal of the valley gutters, while for case 7 the dates coincided with the execution of several maintenance works with government funding. This allowed to identify already three practical considerations that moderate the relation between the response time and the severity, namely (1) the accessibility of the damages, (2) the clustering of interventions as part of larger interventions, and (3) the influence of funding opportunities.

6 IN-DEPTH ANALYSIS OF TWO CASE STUDIES

Two case studies were analysed more in-depth by identifying the chronological sequences based on reports and project documents of interventions. This allowed to establish a time frame and causal connection between the inspection reports and the maintenance interventions. The focus is again on the building's rainwater disposal system.

6.1 *Case 2: Good intents, but errors in execution*

During the in-depth interviews, the owner of case 2 indicated to be concerned for the historical fabric and demonstrated a tendency towards a preventive conservation approach. The average frequency (f) of Monumentenwacht inspections

Table 3. Overview of inspection frequencies, based on data MOWAv.

	Inspection dates	Number inspections	Average time interval	Average frequency
Case 2	1999,2004,2011,2015	4	5.33 years	0.188
Case 3	2001,2004,2009,2011,2015	5	3.5 years	0.286
Case 6	1995,1998,1999,2002,2005,2014	6	3.8 years	0.263
Case 7	1996,1998,2000,2002,2005,2010,2014	7	3 years	0.333

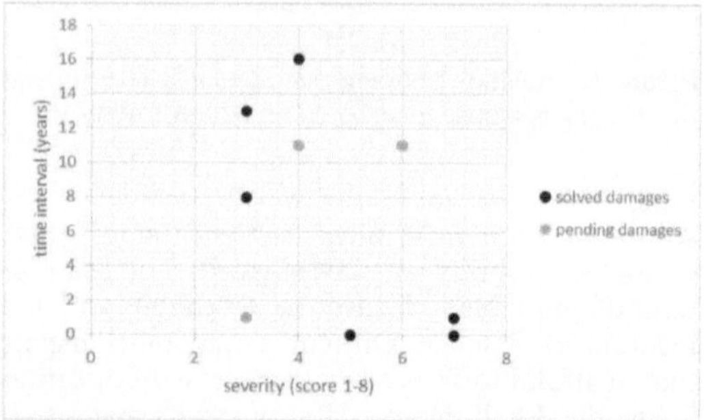

Figure 3. Solved and pending damages, case 2, based on data MOWAv.

is once every 5.33 years (f = 0.188). The average response time is 6.1 years. Only 30% of the detected damages have not yet been solved and the average severity of the detected damages is 4.7. Figure 3 indicates the solved and pending damages for case 2. Wheareas the data in Figures 1 and 2 was based only on the information obtained from the inspection reports, the time intervals in Figure 3 were determined based on inspection reports and project documents.

The good intent of the owner is undone by the inability to control the quality of executed interventions. This is demonstrated in Figure 4 for the main roof covering and its connections. The graph plots the severity of damages with respect to the years of inspection (1999, 2004, 2011, 2015). Spheres represent the roof covering and squares represent the connections of the covering with the masonry. When the spheres or squares are coloured in, infiltrations were detected. Vertical arrows indicate interventions. The roof covering of the flat roof in roofing material, that had already been renewed in 1994, was completely renewed in 1999, based on the recommendations of Monumentenwacht, due to the sloppy placement of the initial covering and the poor execution of connections. After the intervention of 1999, problems were again detected regarding the sloppy placement of the roof covering, resulting in superficial cracks, mechanical damage, folding of the borders, loosening of connections and loosening of seams. In 2011, Monumentenwacht detected leakages (see Figure 5), which were repaired in 2012. This reduced the severity for the connections from 7 to 3, since the damage type changed from loosening of the connection (score 3) to disintegration (score 1) and the condition changed from moderate (3) to tolerable (2). Similarly the severity for the roofing material reduced from 6 to 4.

6.2 *Case 5: Durability of initial design choices and repairs*

During the in-depth interviews, the owner of case 5 demonstrated a thorough understanding of the building's historical evolution and significance. The concern for the historical fabric was however strongly related to the building's interiors. This might be related to the intense restoration that the dwelling was subjected to in the period 1985–1986, due to its poor state of preservation after purchasing it a few years before. The intense restoration required a large investment, which led to some cuts in the expenses by applying less expensive but also less durable materials. A shift in the choice of materials was noted after the listing in 2005. Many of those initial choices were rectified over the years, with the aid of government funding. The average frequency (f) of Monumentenwacht inspections is once every 3.5 years (f = 0.286). The average response time is 10.58 years. About 47% of the detected damages were not yet solved and the average severity of the detected damages is 5.1. A renewal of the roofing material is planned in 2017, so it is expected that most of the pending damages will be resolved in the next inspection report.

The compiling of a chronological sequence for the gutters and the roof covering, enabled to understand more of the owner's maintenance approach. The chronological sequence was reconstructed based on the inspection reports, the project documents and interviews with the owner. The project documents are related to the initial restoration of 1985–1986 and the renewal of the gutters in 2014.

Regarding the gutters, the invoices of the 1985–1986 restoration project indicate that initially it was foreseen to line the timber gutters with zinc. On the 6th of December 1985, the contractor proposed a reduced price for realizing the lining of the gutter bottoms in asphalt, while the elevations are lined

Table 4. Overview of damages solved within 5 years time, based on data MOWAv.

Case	Damage type	Condition	Severity	Urgency	Time interval	Description	Accessibility
2	2	3	5	Short term	5	sloppy placement roofing material flat roof	Easy
2	3	4	7	Short term	5	Repair tiles balcony	Easy
2	2	3	5	Short term	5	Repair connections roofing material flat roof	Easy
2	3	4	7	Urgent	4	Repair leak roofing material flat roof	Easy
5	2	3	5	Short term	3	Repair loosened welds flat roof	Easy
5	3	3	6	Short term	3	Repair leak eaves gutters	Medium
5	4	4	8	Urgent	2	Shifted tiles near valley gutters	Difficult
7	4	4	8	Urgent	3	Replace missing soaker	Difficult
7	1	4	5	Urgent	3	Repair mortar connection	Difficult
7	1	3	4	Short term	3	Renewal gutters	Difficult
7	3	4	7	Urgent	5	Replacement damaged and missing tiles	Difficult
7	1	2	3	Long term	2	Repointing connection raised gables	Difficult
7	2	2	4	No recommendation	2	Repointing connection raised gables	Difficult

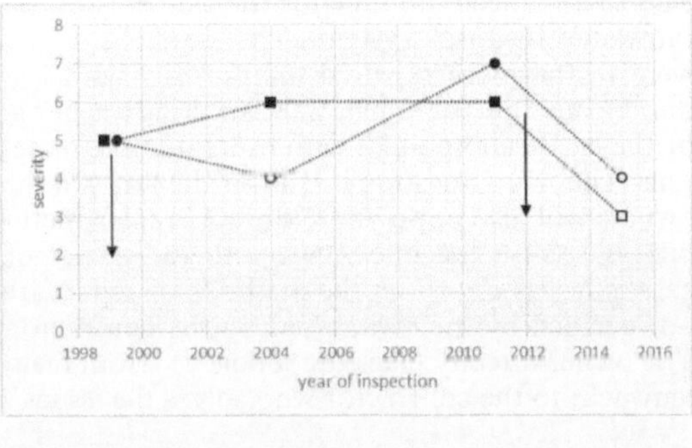

Figure 4. Chronological sequence severity of damage with respect to years of inspection, based on data MOWAv.

Figure 5. Loosening of connections, 2011, Picture by MOWAv.

with zinc, that is nailed to the gutter's timber frame. The initial inspection report of Monumentenwacht dates back to 2001, so 15 years after the execution. The report indicates that the applied materials are less durable and refers to a number of problems with the execution. Firstly, the lining of the gutter bottoms in asphalt tends to fold up, a problem that already caused leakages of which some had been repaired in the past. The report of 2001 mentions previous repairs of leakages. Secondly, the lining of the elevations in zinc demonstrated problems of loosening of the soldered seams, which again caused infiltration. Furthermore, there is a technical design issue that aggravates the situation: some of the downpipes were eliminated during the restoration works of 1985–1986, which resulted in a reduced slope of the gutter towards different exits. Water was therefore less efficiently drained and rainwater would stand in the gutter. The combined issues led to decay of the gutter's timber structure, which in 2005 is mentioned to have caused a subsiding of the gutter structure, again resulting in more standing water in the gutters. Some of the problems were temporary repaired with often non-durable

Figure 6. Gutters in asphalt, 2011, Picture by MOWAv.

Figure 7. Uncovering of roof sheathing and pan laths due to shifting of tiles, 2015, Picture by MOWAv.

techniques and materials. For example, the leakages of the gutter bottoms in asphalt were repaired by gluing lead flashings onto the asphalt, but the glue would loosen, leading again to leakages. Those leakages were then again repaired with more gluing. This temporary repairs continue until the replacement of all the gutters with new zinc linings in 2014.

Regarding the roof covering in ceramic tiles, there were four design issues regarding its initial placement in 1986: (1) a sloppy placement of the ceramic tiles, (2) the use of different tile sizes, (3) the lack of a double tiling batten to support the lower row of tiles, (4) the durability of the applied materials. The first two issues were only mentioned for the first time by Monumentenwacht in 2011, when it was clear that the sloppy placement lead to a bad connection between the tiles and the shifting of the tiles. Some of the shifted tiles were repositioned using building foam or they were repaired with adhesive tape. The lack of the double tiling batten led to the shifting of the lower row of tiles and the uncovering of the ends of the pan laths and the roof sheathing, which accelerated their deterioration. Strangely this design issues was not resolved during the placement of new gutters in 2014. Finally, the durability of the applied materials, is related to the reuse of existing tiles for the restoration of 1986. It was common practice at that time to use recycled old tiles from other buildings for restoration projects, to give the roof coverings an authentic character. Already in 2014, only 28 years after the placement of the tiles, the reports of Monumentenwacht indicate the first signs of deterioration of the roof tiles. The initial superficial deteriorations increase the tiles' porosity, which leads to a higher rate of water absorption and increases the chance of mechanical damage due to frost. Another issue regarding durability of applied materials is the use of cement based mortars for connections of ridges and raised gables, which leads to deterioration of the hard mortars and local infiltration. The roof coverings are planned to be replaced with government funding in 2017, only 30 years after their initial placement.

This case demonstrated how design choices lead to a reduced life time for the elements of the rainwater disposal system. The detected issues were furthermore repaired temporarily with less durable methods and materials, awaiting a renewal of the entire component with more durable materials. The main question is if after the renewal the owners will also apply more durable repair methods. Based on the interviews with the owner of case 5, it was clear that the methods of repair are not decided by the owner, but by the contractor. The owner directly takes the report of Monumentenwacht to the contractor who solves the issues.

7 DISCUSSION

This initial analysis of the owners' maintenance objectives enabled to identify that even though all seven private owners are member of Monumentenwacht, only five of them are active participants in the Monumentenwacht model. The motivation why these owners participate in the Monumentenwacht model could however not be clarified based on the identification of six maintenance objectives.

Four of the case studies were investigated more in detail, in order to understand how the owners operate within the Monumentenwacht model. It was found that the inspection reports inform the owners of the condition and allow them to prioritize maintenance interventions. The tendencies demonstrate that owners take this into account. Six factors were identified that influence this direct relationship between an understanding in the severity of the damages and the responses of the owners: (1) the accessibility of the location of damage, (2) the clustering of interventions, (3) the availability of funding, (4) the lack of quality control during the execution of works, (5) the durability of the initial design choices, (6) the availability, knowledge and skills of the contractor.

Due to the limited number of cases, this initial analysis remains a sample study of local practice. The study demonstrated that inspection reports can provide insights in the owners' responses. Therefore, ongoing research entails a statistical analysis of a larger set of cases based on inspection reports in order to understand if similar tendencies can be detected.

The in-depth case study analysis demonstrated the limitations of a mere investigation of the reports and the added value of the applied mixed method approach. Therefore, the same method will be applied to investigate a similar set of cases in a rural context. Finally, since the in-depth investigation of case 2 and case 5 have demonstrated the importance of quality control, design choices and durability of interventions, this case study research will also be integrated within ongoing research on quality management of interventions.

ACKNOWLEDGMENTS

The authors gratefully acknowledge the support of the Agency for Innovation by Science and Technology (IWT) for the Ph.d. grant of Nathalie Van Roy and the JPI Heritage Plus program for the support of the European Project CHANGES (Changes in Cultural Heritage Activities: New Goals and Benefits for Economy and Society).

REFERENCES

CHANGES (2017). Changes in cultural heritage activities: New goals and benefits for economy and society, www.changes project.eu, last consulted 05.07.2017.

Dann, N. (2013). Preventive conservation maintenance in england: Constraints, opportunities and lessons. In K. Van Balen and A. Vandesande (Eds.), *Reflections on Preventive Conservation, Maintenance and Monitoring of Monuments and Sites*, Leuven, Belgium, pp. 95–100. Acco.

Dann, N. & T. Cantell (2005). Maintenance: From philosophy to practice. *Journal of Architectural Conservation 11 (1)*, 42–54.

Dann, N. & S. Wood (2004). Tensions and omissions in maintenance management advice for historic buildings. *Structural Survey 22 (3)*, 138–147.

Dann, N., D. Worthing, & S. Bond (1999). Conservation maintenance management establishing a research agenda. *Structural Survey 17 (3)*, 143–153.

Della Torre, S. (2013). Planned conservation and local development processes: the key role of intellectual capital. In K. Van Balen and A. Vandesande (Eds.), *Reflections on Preventive Conservation, Maintenance and Monitoring of Monuments and Sites*, Leuven, Belgium, pp. 123–127. Acco.

Elena Charola, A., M. Henriques Fernando, & L. Magadan Marcelo (2015). The relevance of maintenance and monitoring in architectural conservation. *Restoration of Buildings and Monuments 21(1)*, 1–8.

Feilden, B. & J. Jokilehto (1998). *Management and guidelines for World Heritage Sites*. Rome: International Centre for the Study of the Preservation and Restoration of Cultural Property.

Forster, A. & B. Kayan (2009). Maintenance for historic buildings: A current perspective. *Structural Survey 27(3)*, 210–229.

Glaser, B.G. & A.L. Strauss (1967). *The discovery of grounded theory: strategies for qualitative research*. Chicago: Aldine Publishing Co.

ICOMOS (1999). The burra charter.

ICOMOS (2003). Iscarsah principles for the analysis, conservation and structural restoration of architectural heritage.

Marconi, P. (1984). *Arte e cultura della manutenzione dei monumenti (E: Art and culture of maintenance of monuments)*. Bari: Laterza.

Moioli, R. (2013). Non-use values and the assessment of performance of preventive and planned conservation. In K. Van Balen and A. Vandesande (Eds.), *Reflections on Preventive Conservation, Maintenance and Monitoring of Monuments and Sites*, Leuven, Belgium, pp. 95–100. Acco.

Mortelmans, D. (2013). *Handboek kwalitatieve onderzoeksmethoden (E: Manual on qualitative research methods)*. Leuven: Acco.

Van Balen, K. (1988). Stabiliteitsherstel in monumentenzorg (e: structural consolidation in monument conservation). In R. Lemaire and K. Van Balen (Eds.), *Stable-Unstable?*, Leuven, Belgium, pp. 15–27. Leuven University Press.

Van Balen, K. (2015). Preventive conservation of historic buildings. *Restoration of Buildings and Monuments 21(2–3)*, 99–104.

Van Balen, K. & A. Vandesande (2013). Introduction. In K. Van Balen and A. Vandesande (Eds.), *Reflections on Preventive Conservation, Maintenance and Monitoring of Monuments and Sites*, Leuven, Belgium, pp. ii–vi. Acco.

Vandesande, A. (2017). *Preventive Conservation Strategy for built heritage aimed at sustainable management and local development*. Leuven: PhD dissertation.

Verstrynge, E., L. Schueremans, & P. Smars (2012). Controlled intervention: Monitoring the dismantlement and reconstruction of the flying buttresses of two gothic churches. *International Journal of Architectural Heritage 6*, 689–708.

Yin, R. (2014). *Case study research: design and methods*. London: SAGE.

Innovative Built Heritage Models – Van Balen & Vandesande (Eds)
© 2018 Taylor & Francis Group, London, ISBN 978-1-138-49861-7

HeritageCARE: Preventive conservation of built cultural heritage in the South-West Europe

L.F. Ramos, M.G. Masciotta, M.J. Morais, M. Azenha, T. Ferreira,
E.B. Pereira & P.B. Lourenço
Department of Civil Engineering, ISISE, University of Minho, Guimarães, Portugal

ABSTRACT: Currently, no systematic policy for the preventive conservation of built cultural heritage exists in South-West Europe. Existing approaches for inspection, diagnosis, monitoring and curative conservation are normally intermittent, unplanned, overpriced and lacking in methodical strategy. The available financial resources are scarce and they are mostly addressed to listed buildings. Besides, owners often conceal an inborn reluctance to invest in preventive conservation and maintenance programs.

In the light of these considerations, the HeritageCARE project has recently been launched with the purpose of unfolding an integrated and sustainable strategy for the preventive conservation of built cultural heritage within South-West Europe. This project involves 3 Countries (Portugal, Spain and France), 8 beneficiary partners and 11 associated partners. The main aim of the paper is to present the key aspects of the project, its objectives and challenges, and to outline the multilevel methodology that will be implemented and validated through 60 case-study structures.

1 INTRODUCTION

With a population of 76.8 million inhabitants and an average density of 97 inhabitants/km², the Southwest Europe—Portugal, Spain and Southwest France (Interreg-Sudoe 2017), hereafter called SUDOE, boasts approximately 31,163 heritage listed buildings (Fig. 1), being unknown the number of non-listed buildings with architectural or historic interest.

The cultural identity of the SUDOE territory is deeply reflected into the construction systems spread all over this area. Although different technical solutions and morphological/typological portrayals are found in each of the SUDOE regions, common types of construction systems, damage and deterioration processes can be identified (HeritageCare 2017a).

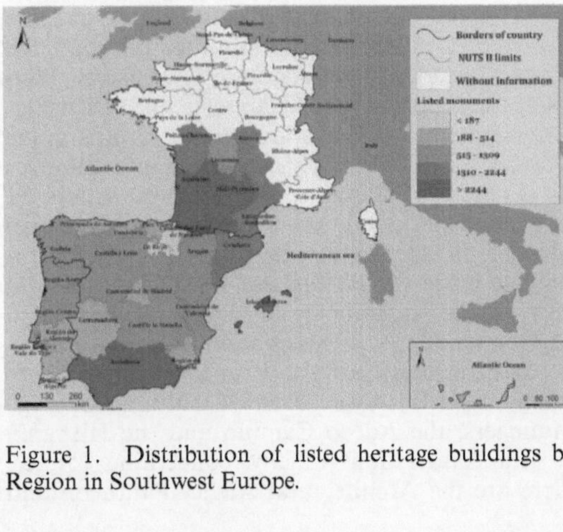

Figure 1. Distribution of listed heritage buildings by Region in Southwest Europe.

In terms of heritage preservation, the driving principles of the SUDOE countries are very well aligned with the UNESCO methodology and recommendations (UNESCO 1954, 1970, 1972, 2001, 2003 & 2005). Nevertheless, a systematic and integrated approach for preventive conservation of built cultural heritage does not exist in these regions. This is not only due to the lack of knowledge about good maintenance and preventive conservation practices, but also to the lack of financial resources, which hinders owners from undertaking conservation actions on their properties.

Current strategies mainly focus on curative conservation and restorative treatments, often carried out when the severity and visual impact of damage are of such an extent that urgent remedial measures become imperative. This reactive approach leads to overpriced interventions and considerable investments of societal and financial resources. In most cases, national funding is insufficient or addressed to listed buildings and substantial interventions, thus the great majority of built heritage is at risk. Another important aspect is the absence of appropriate maintenance and proactive preventive conservation plans to help owners minimize damage and deterioration processes on their buildings in the medium/long term. Last, but not least, is the scepticism of building managers/owners to invest in preventive conservation programs due to the lack of information about the real benefits associated with regular inspections and prevention measures (HeritageCare 2017a).

At government level, each SUDOE country has its own institutions designated to the preventive conservation of cultural heritage. However, no integrated and sustainable strategy able to involve all different actors working in the fields of inspection, monitoring and preventive conservation of built environment has been implemented up to now. Each country acts on its own by following its own actuation strategy and applying its own management tools.

Figure 2. Actors involved in the HeritageCARE project.

2 THE HERITAGECARE PROJECT

2.1 *Focus and objectives*

The HeritageCARE project—monitoring and preventive conservation of historic and cultural heritage—arises in response to the need of a systematic and sustainable approach for the preventive conservation of built heritage in Southwest Europe. A funding of 1,686,282.82 € has been allocated by the European Regional Development Fund (ERDF) within the Interreg-SUDOE program to finance this three-year project, which has started in September 2016. For the purpose of this paper, the definition of 'cultural heritage' adopted in the UNESCO Convention of 1972 shall be considered (Jokilehto 2005).

Driven by the principle "*prevention is better than cure*", the focus of the HeritageCARE project is to give rise to a joint strategy based on a multi-level system of services for the systematic inspection, diagnosis, monitoring and management of heritage assets—whether listed or not—in the SUDOE territory. Beyond traditional tools, such a strategy will leverage the latest advances in digitization and smart technologies to keep abreast of the times and further enhance the quality of the services provided.

Among the main objectives that HeritageCARE envisages to accomplish in the short/medium term are the creation of a database for buildings/assets management, information exchange, and the standardization of methods and tools for maintenance and preventive conservation of heritage buildings/assets. In this regard, guidelines and rules of "*good practice*" will be developed during the project. In the long-term, the HeritageCARE mission is to raise public awareness about the societal and economic benefits associated with the adoption of preventive conservation measures as well as to engage society in the conservation process of historic assets by involving all possible actors working in heritage-related fields (Fig. 2), such as owners and managers of historic buildings; tourists; heritage management entities; national and regional authorities; construction, rehabilitation and inspection sectors; universities and research centres.

The ultimate goal and challenge of the project is the creation of a non-profit self-sustaining entity that will keep pursuing the HeritageCARE mission once the project is concluded.

2.2 *The Consortium*

The project Consortium is composed of 8 beneficiary partners, counting a total number of 27 collaborators and researchers distributed among the three partner countries, i.e. Portugal, Spain and France. Beyond the University of Minho, leader and coordinator of the project, there are the Regional Direction of North Culture and the Centre of Graphic Computation for Portugal; the Santa Maria La Real Foundation of Historic Heritage, the University of Salamanca and the Andalusian Institute of Historic Heritage for Spain; the Universities of Clermont-Auvergne and Limoges for France. Besides, the Consortium counts 11 associated partners, whose support will be indispensable for the success the project. As for Portugal, there are the Portuguese Association of Old Houses, the Archdiocese of Braga and the Guild of Companies for Conservation and Restoration of Architectural Heritage. Regarding Spain, there are the Spanish Association of Cultural Heritage Managers, the Adeco Camino and the Hispania Nostra Association. Finally, concerning France, there are the Architectural Atelier Pantheon and the Architecture Office Pascal Parmantier—Heritage Architect.

The HeritageCARE headquarters is based in Guimarães, at the Institute of Science and Innovation for Bio-Sustainability (IB-S) of the University of Minho, leader partner of the project.

3 THE METHODOLOGY

The methodology through which HeritageCARE will operate in the three countries is soundly aligned with CEN standards (EN 15898, EN 16095, EN 16096) and relies on a system of services for the periodic inspection and monitoring of historic buildings supported by advanced technologies. This system of services is organised in three levels (Fig. 3). Each service level includes the previous one, thereby featuring different costs per different levels. The choice of the service level depends on the owners' needs and financial resources. However, HeritageCARE will try to keep all three service levels as affordable as possible to help owners/managers embark on the conservation process of their properties.

3.1 *Service Level 1 – StandardCare*

Although less complex and demanding than others, Service Level 1 (SL1) is considered the most important of HeritageCARE service levels. Simple, feasible and low-cost, SL1 shall allow a rapid condition screening of the conservation status of the heritage buildings along with their integrated assets. The condition survey is carried out through on-site inspections and is aimed at identifying building pathologies and deterioration processes in due time.

Prior to inspection, a historic analysis of the building is made to get acquainted with previous structural interventions, maintenance works, functional changes over time, and so forth. Then, qualified teams of professionals perform detailed on-site inspections with the support of specific fillable forms/e-forms. The overall condition of the construction is assessed based on a grading scale system (Table 1) that associates a weighted index to each of the inspected items (HeritageCare 2017b).

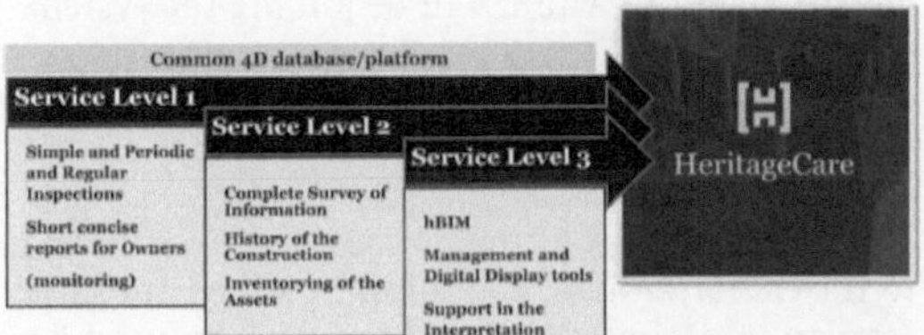

Figure 3. Schematic summary of HeritageCARE methodology.

According to this criterion, the need for urgent remedial measures or additional inspection and diagnosis works is automatically detected.

Based on the inspection outcome, HeritageCARE will elaborate a short easy-to-read report with advices and recommendations for owners, pointing out all necessary conservation and maintenance actions that they should undertake in the short, medium and long term to prevent further decay. Depending on the type and conservation status of the object, inspections will be repeated at regular intervals, e.g. every 2–3 years, thereby allowing to compare the health condition of the building over time.

Table 1. HeritageCARE grading system for condition and risk classification.

Class No.	Condition classification	Symptoms	Urgency risk classification	Comments
0	Good	No symptoms	Long term	No immediate actions required | Preventive monitoring is necessary
1	Fair	Minor Symptoms	Medium term	The condition of the fabric is not perfect but does not need immediate action | Monitoring is necessary to prevent further decay
2	Poor	Moderately Strong Symptoms	Short term	The condition of the fabric is such that it needs timely repair or additional inspection and diagnosis work
3	Bad	Major Symptoms	Urgent and Immediate	Urgent repair is necessary | Urgent additional inspection and diagnosis work
NA/NI	Not Accessible	Parts not (safely) accessible	Not Inspected	Parts that are 'not inspected' are either not (safely) accessible for the building inspectors or not visible

Object information, inspection data and reports will be progressively stored in a 4D database to keep track of the conservation status of all inspected buildings. This database will be used to feed a web-based platform for HeritageCARE subscribers to easily access and retrieve the stored information (for privacy's sake access to contents will be restricted depending on the user authorization). In case of serious problems, both owners and national Heritage authorities will be alerted through the system. If necessary, owners will be also advised to contact experts in order to perform in-depth inspection and diagnosis works, and to plan a prompt intervention.

3.2 *Service Level 2 – PlusCare*

Service Level 2 (SL2) is conceived to increase and further detail the level of information of the inspected buildings and related indoor assets, like altars, tapestries and paintings to cite a few instances, integrating and complementing the information collected in SL1. To this end, advanced techniques and equipment might be required, such as geomatics and associated tools for 3D recording (e.g. UAV photogrammetry, laser scanner, backpack mapping). Nevertheless, HeritageCARE commits to keeping SL2 fieldwork as convenient and viable as possible, and so the final products.

The main outputs of this level are the 3D modelling and virtualization of the inspected heritage structure, with the incorporation of all information acquired up to this stage (beyond SL1 information, there might be additional documentation from the inventory of integrated assets or data from structural monitoring activities and ND testing). These models will be also used to produce 2D technical drawings of the construction, i.e. plans, sections and elevations.

The data and outputs generated in SL2 will be uploaded, organized and managed in the database, thus allowing HeritageCARE subscribers to access them anytime through the afore-mentioned web platform and eventually use this material for tourism purposes, heritage promotion and fruition, etc.

3.3 *Service Level 3 – TotalCare*

Service Level 3 (SL3) is devoted to the integration and management of information obtained from former levels into the HeritageCARE database, by leveraging the *Building Information Modelling* (BIM) with application to heritage structures—hence the acronym hBIM (Arayici et al., 2017). Digital models of the inspected historic buildings are built, accessed and manipulated through this intelligent 3D model-based process, including building constituent materials, integrated assets, and costs associated with maintenance plans and management operations. In other words, BIM environment enables faster information sharing, more accurate resolutions, visual representation of potential issues, better integration between preventive conservation actions and management systems, and much more.

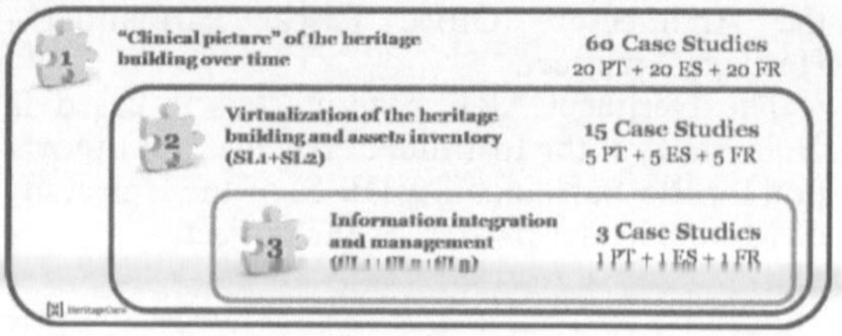

Figure 4. Distribution of HeritageCARE case studies by service level.

Supported by hBIM, SL3 will allow HeritageCARE to streamline the conservation and management process of complex and singular heritage constructions, resulting of great help for owners and managers of monumental fabrics. Fulcrum of SL3 is to guarantee the interoperability between hBIM models and the database of HeritageCARE at SL1/SL2.

3.4 *Implementation and validation*

During the project, the HeritageCARE methodology presented in the previous sections will be implemented and validated through real case-study structures to be chosen among the great variety of heritage buildings spread over the SUDOE territory. Note that these case studies will be selected so as to encompass the most representative structural typologies identified in the SUDOE territory.

Service Level 1 will be applied to 60 case studies, namely 20 buildings per partner country. Out of these 60 case studies, 15 will be selected for the implementation of Service Level 2, counting 5 buildings per country. Finally, considering the higher complexity of the services when raising the level of the methodology, Service Level 3 will be exclusively applied to one building per country, resulting in a total number of 3 case studies (Fig. 4). During the project time frame, the services provided by HeritageCARE will have no cost or charge for those whom will be joining the system.

4 ONGOING PROJECT ACTIVITIES

The initial months of the project have been devoted to the elaboration of a first report aimed at characterizing and contextualizing the built cultural heritage present over the SUDOE territory as well as shedding light on the current status of preventive conservation in the three SUDOE countries. The scope of the report was to gather and organise all information necessary to embark on the activities established for the three service levels and to outline the best actuation strategy in terms of conservation measures and recommendations. Online questionnaires have been sent to owners and managers of the historic and cultural buildings belonging to the SUDOE territory to get a better insight into the *status quo* of such properties, especially regarding use, state of conservation and major maintenance problems experienced. The information so collected is now being exploited for the development of *ad hoc* inspection tools, such as interactive mobile apps linked to a damage atlas for assisting HeritageCARE inspectors in the characterization of the types of damage affecting the buildings and in the identification of possible causes and consequences of damage. It is worth noting that an efficient preventive conservation program can only be planned on the basis of a correct and thorough diagnosis.

The afore-mentioned report provided an overview of the most common construction systems, types of damage and deterioration processes that are found within the SUDOE territory. Each region features different morphological and typological portrayals, but dwelling, religious and military constructions are the most recurrent structural typologies in Southwest Europe. In what concerns the construction systems, five main types have been identified, namely masonry, rammed earth, timber, iron and concrete constructions. For each system, the relevant construction components have been classified and grouped based on their structural role within the building, distinguishing among structural system, building envelope, partitions and installations (HeritageCare 2017a).

As for the integrated objects, it was deemed more appropriate to categorize them depending either on the type of material (stone, wood, gold, silver, copper, iron, ceramic, textile and others) or on the type of object (paintings, frescos, sculptures, altarpieces, furnishings, glazed tiles or other kinds of movable objects, such as weapons, books, fabrics, jewellery and the like).

Definition and illustrations of recurrent types of damage and deterioration processes for both construction systems and integrated objects are currently being organized in the form of a damage atlas (HeritageCare 2017a, MDCS 2017, Monumentenwacht 2017, ICOMOS 2008), where potential causes, consequences and correlations between different types of damage are identified and ordered based on the affected materials and construction elements as well as according to possible prevention and mitigation actions. For obvious reasons, besides English, the damage atlas is being translated in the three SUDOE languages, viz. Portuguese, Spanish and French.

At the present time, the HeritageCARE Consortium is detailing the key steps of the methodology for preventive conservation of historic and cultural buildings. The technical requirements and specifications for the tools to be used in each service level are also being defined along with the layout and contents of the outputs to deliver to owners and stakeholders (Fig. 5).

These activities are resulting into two additional reports which will mark the conclusion of the first year of the project (HeritageCare 2017b,c). Even though still in progress, the methodology is already being tested on a few case studies (beginning from Service Level 1) so as to evaluate its pros and cons, and opt for a refinement in case of necessity. A critical appraisal is fundamental at this stage, before proceeding with the inspection of all

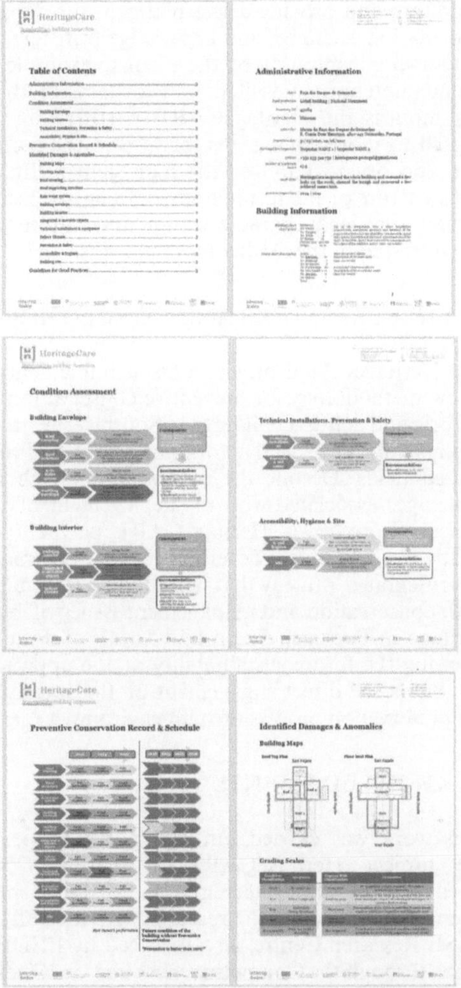

Figure 5. Example of layout for owner's report.

selected case-study structures and the standardization of methods & tools for the development of guidelines and rules of *"good practice"*.

Transversal tasks related to the management and communication of the HeritageCARE project are being run as well, paying particular attention to the dissemination and promotion of the objectives, activities and progressive achievements of the project among the desired target audiences. To this purpose, both traditional media channels and new Internet communication vehicles, like webpages and social media platforms, are being used so as to reach out as many people as possible and to raise public awareness of the benefits of regular inspections and maintenance programs in the conservation process of built cultural heritage. Target-oriented seminars, forums, workshops and training schools will be organised to present the multi-level methodology through which HeritageCARE is operating and provide a live platform for debate, information exchange and knowledge transfer.

Drawing inspiration by the Flemish example of Monumentenwacht visited in February, in these last months the HeritageCARE partnership has started to move some steps forward in the definition and organization of the management/financial structure of the future non-profit entity. Issues related to its legal characterization and statute will be addressed in the forthcoming months.

5 FINAL REMARKS AND CONCLUSIONS

The HeritageCARE project aims at implementing a new methodology for preventive conservation of historic and cultural buildings in Southwest Europe by involving all societal actors working in heritage-related fields. Despite the economic and cultural challenges associated with the achievement of the envisaged objectives, HeritageCARE partners are jointly working for the development of a systematic and integrated strategy that will allow to streamline both conservation and management issues of built heritage in Portugal, Spain and South of France, ensuring the future sustainability of the approach as well as the direct engagement of the society in the preservation process of built environment.

ACKNOWLEDGMENTS

This work was carried out within the scope of the project HeritageCARE (SOE1/P5/P0258), co-funded by the Interreg-Sudoe/FEDER program and included in the research activities of the ISISE Research Centre, also financed by FEDER funds through the Competitiveness Factors Operational Program—COMPETE and by national funds through FCT—Foundation for Science and Technology within the scope of the project POCI-01-0145-FEDER-007633.

REFERENCES

Arayici, Y., Counsell, J., Mahdjoubi, L., Nagy, G., Hawas, S. & Dweidar, K (Ed.) (2017). Heritage Building Information Modelling, 1st Edition, Routledge, ISBN-13: 978-1138645684, ISBN-10: 1138645680.

EN 15898:2011 Conservation of cultural property—Main general terms and definitions.

EN 16095:2012 Conservation of cultural property—Condition recording for movable cultural heritage.

EN 16096:2012 Conservation of cultural property—Condition survey and report of built cultural heritage.

HeritageCare. (2017a). Survey of construction systems, type of damages and deterioration processes within the SUDOE territory. Unpublished document, *HeritageCARE* (Report of the Project Activity 1.1).

HeritageCare. (2017b). Technical requirements for the tools to develop in the HeritageCare project. Unpublished document, *HeritageCARE* (Report of the Project Activity 1.2).

HeritageCare. (2017c). General methodology for the preventive conservation of cultural heritage buildings. Unpublished document, *HeritageCARE* (Report of the Project Activity 1.3).

ICOMOS. (2008). Illustrated Glossary on Stone Deterioration Patterns. *Monuments and sites* (xv).

Interreg-Sudoe. (2017). https://www.interreg-sudoe.eu/gbr/programme/what-is-interreg-sudoe-programme.

Jokilehto, J. (2005). Definition of cultural heritage: References to documents in history. ICCROM Working Group 'Heritage and Society'.

MDCS. (2017). The Monument Diagnosis and Conservation System, "mdcs.monumentenkennis.nl".

Monumentenwacht The Netherlands, 2017. "www.monumentenwacht.nl".

UNESCO. (1954). Convention for the Protection of Cultural Property in the Event of Armed Conflict with Regulations for the Execution of the Convention, United Nations Educational, Scientific and Cultural Organization, Hague Convention.

UNESCO. (1970). Convention on the Means of Prohibiting and Preventing the Illicit Import, Export and Transfer of Ownership of Cultural Property, Paris Convention.

UNESCO. (1972). Convention concerning the Protection of the World Cultural and Natural Heritage, United Nations Educational, Scientific and Cultural Organization, The General Conference of the UNESCO, Paris.

UNESCO. (2001). Convention of the Protection of Underwater Cultural Heritage, United Nations Educational, Scientific and Cultural Organization, The General Conference of the UNESCO, Paris.

UNESCO. (2003). Convention for the Safeguarding of the Intangible Cultural Heritage, United Nations Educational, Scientific and Cultural Organization, The General Conference of the UNESCO, Paris.

UNESCO. (2005). Convention on the Protection and Promotion of the Diversity of Cultural Expressions, United Nations Educational, Scientific and Cultural Organization, The General Conference of the UNESCO, Paris.

Case studies of change

Innovative Built Heritage Models – Van Balen & Vandesande (Eds)
© 2018 Taylor & Francis Group, London, ISBN 978-1-138-49861-7

Conservation dynamics. A new paradigm for cultural heritage in times of economic changes

C. Ost
ICHEC Brussels Management School and Raymond Lemaire International Centre for Conservation, KU Leuven University, Belgium

F. Carpentier
Exhibition Coordinator, Centre for Fine Arts Brussels (BOZAR), Belgium

ABSTRACT: This paper aims to develop a new paradigm for cultural heritage in times of economic and social transitions. The analysis derives from a key-note conference delivered at the KU Leuven (Ost, 2017), and from complementary architectural-cultural reflections on the relationship between heritage values and program renewal in conservation narratives (Carpentier, 2014). The authors analyze the post-WWII social and economic context that presented culture as a fresh domain for modern analysis, and a form of resilience towards the international expansion of uniform ways of life. Economic conditions have remained important for heritage conservation, and sometimes a catalyst for changes in practices for heritage management and financing. The paper selects some major trends and structural changes that characterize the current long-wave context of economic transition, and proposes a new paradigm aimed to better integrate cultural heritage in our society today.

1 INTRODUCTION

Conservation theory and policy of the 20th century mainly reflect the attempts towards a definition of universal ethical codes related to the conservation of tangible qualities of cultural heritage. While the Venice Charter of 1964 was an effort to integrate the conservation of monuments and sites within the dynamic architectural pace in post-war Europe, the Nara Document on Authenticity critically questioned the 'top-down' guidelines of the Venice Charter to include alternative views on authenticity based on various cultural traditions and practices.

This paradigm shift towards the intangible features related to the conservation of monuments and sites, already visible in the spirit of the Nara document, fully started in 2008 when the 2003 Convention for the Safeguarding of the Intangible Cultural Heritage took effect, creating a List of Intangible Cultural Heritage, independent from physical artifacts.

While the paradigm shift towards the intangible has been essential to widen the scope of cultural heritage practices, it also presents a danger for heritage to be used as an ideological tool. The spontaneous evolution of contemporary cultural practices seem to be overruled by nostalgic, artificial reflexes to invest in and safeguard practices from the past. There is a danger of overthinking, over-intellectualizing everything. As Andrea Bruno puts it: "In the Middle Ages the obsolete Roman amphitheaters were converted to new uses; there were no problems of the conservation of past memories as evidence of history and civilization. Utilitarianism and urgency led to total, unscrupulous but extremely forceful reappropriations (Bruno, 1996)."

In this paper, we therefore plead for a new paradigm shift, one that transcends the tangible and intangible, and does not focus on the *a priori* conservation of cultural heritage, but starts from the cultural, social, and economic needs of contemporary society. By introducing a new paradigm, the ways we have been looking at cultural heritage until now can be reversed. The conservation of a monument, site, or cultural heritage practice is not the starting point anymore, but may become an effective resource for sustainable development.

2 HERITAGE ECONOMICS: A FAST-GROWING FIELD

Heritage economics is a fairly new branch of economics, derived from both cultural economics and environmental economics during the late 1960s. Cultural economics has brought to the analysis of cultural heritage the specificity of its object, being the analysis of a real estate good which offers to its users and to the community as a whole cultural services that are almost similar to the provision of public goods and services (Throsby, 2001). On the other hand, environmental economics provided a set of tools and methodologies that contribute to measure cultural reality with its economic and social components, which often escaped the logic of the market (Baumol & Oates, 1988).

The first academic event in cultural economics was held in Edinburgh in 1979, as the international conference on cultural economics organized by Professor William Hendon from the University of Akron, Ohio, also founder of the Journal of Cultural Economics in 1975. It is not surprising that such academic interest developed during the 1970s, a decade known for major economic changes. The end of the Golden Sixties, with a debate on consumerism, coupled with technological challenges and expanding globalization, has caused major uncertainty about the future economy.

This context presented culture as a fresh domain for modern economic analysis (measuring the immeasurable), and also as an alternative to the short-term profit-seeking economic model of the post WWII reconstruction era. In other words, culture was seen as a form of resilience towards the international expansion of uniform ways of life, notably in Western urban context (WCCD, 1995). The fact that modern cultural heritage protection surged during the Sixties (culminating with the adoption of the 1972 UNESCO World Heritage Convention), was of course the result of the international concern for heritage preservation that emerged after WWII, but was also catalyzed by the urgency to protect heritage from modern economic development.

At the same time, consumerism was accompanied by the emergence of a Western leisure and recreational society, which started to turn into mass-tourism development because of the expansion of mobility and international trade. Cultural heritage benefited and suffered simultaneously from economic growth, which is probably still the case today, confirming that heritage conservation and economics may experience similar transformations.

That context stimulated the production of research aimed to show evidence of the economic and social impacts provided by heritage conservation, measuring cultural return from heritage conservation in terms of jobs, income, growth and welfare (Getty Conservation Institute, 1999). The perspective of cultural heritage considered as an economic resource that could contribute to mitigate impacts from economic recessions flourished faster than expected, and results today in robust scientific advancement of literature in heritage economics, mainstreaming the long-term and sustainable economic impact of cultural heritage on society as the main narrative to unlock public and private funding for the conservation of cultural heritage.

3 ECONOMIC CONDITIONS: BETWEEN PUBLIC AND PRIVATE FUNDING

Over the years economic conditions have remained important for heritage conservation, and an influential factor in cultural decisions. Public policies and financial priorities have always been impacting conservation, since allocation of public resources into the cultural sector follows the ups and downs of economic conditions. Economics also influence cultural heritage through the changing behavior of heritage users (occupants, visitors, tourists), whose consumption and investment decisions are a function of income and market prices. Sometimes such economic influence was positive (increase in the number of visitors, public incentives for urban rehabilitation), sometimes such influence was negative (budget cuts, recession, competing urban projects) (Licciardi & Amirtahmasebi, 2012).

The so-called Great Recession of 2008–2009 was a pivotal moment for cultural actors in taking into consideration the urgent issues related to a strong economic crisis. Although the recession was initiated in the US economy (housing bubble and financial crash), economic conditions were impacted all across the world, including in emerging countries. Heritage protection and conservation were hit by the following factors: decrease in tourism, depressed real estate market, less private and public funding, government cuts in cultural programs.

Some European countries were severely hit by the wave of recession and growing uncertainty that followed. In some places, less financial resources meant that conservation programs had to be suspended, that protection of archeological areas were no more ensured with a risk for looting (Georganas, 2013). Private funding became crucial for keeping some heritage properties safeguarded. Although some information suggests that places with heritage properties were more resilient towards bad economic conditions, evidence has still to be shown if some factors (type of heritage, level of protection, location) could contribute to improve resilience for cultural properties when facing an economic threat (Broadbent, 2011).

The recession of 2008–2009 is just an example of short-term cyclical fluctuations ("business cycles") that are the best known and visible patterns for economic growth (on the average, a recession occurs worldwide every 7 to 10 years) (Stock & Watson, 1993). It is assumed that cultural heritage is impacted by such ups and downs (expansions and recessions) in the economy. Typical cyclical movements explain why property prices of cultural heritage rise or fall, why the number of visitors increases or decreases, why public funding may be tight or loose. Cyclical changes may also impact conservation works, through wages, jobs, income, or production costs.

The undisputed success story of cultural heritage conservation since the adoption of the UNESCO Convention on the Protection of Cultural and Natural Heritage in 1972 (more than a thousand inscriptions today across the world), cannot elude the question on the growing difficulty for conserving our entire tangible and intangible heritage, everywhere around the world, in any types of economic conditions.

A substantial number of initiatives were taken to mitigate impacts of the recession on heritage conservation (Riksantikvaren, 2009). Some are urgent measures or short-term adjustments, some are more structural changes, and the result of long lasting reflections and debates among conservation specialists over time. Short-term and long-term considerations are intermingled here, as short or medium-term recessions interconnect with long term economic growth.

From a time perspective, heritage conservation is embedded in a long-term approach that matches the perennial value of heritage, and its transmission to the next generation. But there is also a short-term consideration for values of cultural heritage, mainly in terms of the uses of heritage properties for recreational, tourism, or real estate purposes. Economically speaking, both short-term and long-term issues of heritage are to be taken into consideration.

Research in business cycle theory has shown that short-term fluctuations were embedded in long waves of economic development connected over time by periods of economic and social transitions (Knoop, 2004). Typically, the Great Recession of 2008–2009 is a cyclical downturn which was aggravated by structural factors related to long wave changes. The coincident evolutions of heritage conservation and economic conditions, as explained above, is a typical example for long wave changes. First, a post-WWII reconstruction period followed by the Golden Sixties that accompanied the raising awareness for natural and cultural protection, then, decades of economic and political transitions (up until today) that accompanied the success story of heritage conservation.

Today, uncertainties and structural changes that are part of economic transitions are challenging the success of heritage conservation. The great transition that the world is experiencing is typified by digital revolution and globalization, concern for climate change and environmental issues, and the mainstreaming of sustainable development. As far as cultural heritage is concerned, the adoption of the United Nations Post-2015 Development Goals (UN, 2012) and the Urban Agenda in Quito in 2016 (UNESCO, 2016) provide a new framework for integrating culture, economic, social, and environmental objectives.

In other words, while the recession of 2008–2009 was a catalyst for changes in current practices for heritage management and financing, the long-term perspective of sustainable development that has been reasserted with the recession becomes the real challenge for conservation, and the opportunity for a new paradigm.

4 MAJOR TRENDS AND STRUCTURAL CHANGES: TOWARDS A NEW PARADIGM

Economist Josef Schumpeter contributed to explain the structural factors that shape long waves of economic growth and periods of economic transitions in between. Factors include technological innovations driven by creativity and entrepreneurship, in a process of changes known as creative destruction (Schumpeter, 1934). It is agreed that heritage conservation may contribute to prepare the new technological wave, also that cultural heritage (mostly in its intangible component) is a source of creativity, and that entrepreneurship becomes a condition for local and community heritage management. As provocative as a process of 'creative destruction' can appear when we deal with conserving built heritage, it still can be of help in regard to the current debate on what and how to conserve in the coming decades. As mentioned above, conservation stakeholders (as contemporary architects and artists in general) did not expect threatening changes in economic conditions to reflect on new principles and to make recommendations to improve protection and conservation today. What needs to be emphasized is the fact that changes in heritage conservation are coincident with moving economic and social conditions.

Some major trends and structural changes that characterized the current long-wave context of economic transition may be analyzed in terms of impact on cultural heritage issues:

4.1 *Glocalism*

The "Globus et Locus" Association defines 'glocalism' as the "momentous changes generated by globalization, changes which have resulted in a permanent intertwining of the global and the local dimensions". This "double process of localization of the flows and globalization of the places (Globus et Locus, 2017)" provides new insights into how cultural heritage can be used as a critical tool to counter the forces of globalization while at the same time using it for the fulfillment of local needs.

The integration of multinational corporations within cultural heritage buildings can be seen as an effect of glocalism. The unavoidable (pressure by multinationals to integrate themselves in key locations) is being countered by the local willingness to preserve a building and oblige the multinational to adapt to local regulations. An example of this process is the re-use of the protected Berlin-Pavilion (1957) into a Burger King. Situated in the Hansaviertel in Berlin, the Berlin Pavilion used to be the entrance of the Interbau 57, an international architecture exhibition where Le Corbusier, Walter Gropius, Oscar Niemeyer and other modern architects joined forces to show the world how advanced architecture had become in West-Berlin, and in the West in general. Today a Burger King has been installed in the historical entrance pavilion. While this new function seems obtrusive, it has been carefully integrated with respect for the existing architecture.

4.2 *Urban century*

In continuity with the effects of glocalism, Eric Corijn, Professor of Social and Cultural Geography at the Vrije Universiteit Brussel, has been for many years one of the voices predicting the evaporation of the national level in favor of the global level (international cooperation) on the one hand, and the local level on the other hand. In his "If Mayors Ruled the World: Dysfunctional Nations, Rising Cities" (Yale University Press, 2013) and through the initiative of the Global Parliament of Mayors, Benjamin R. Barber suggests a future where the democratic epicenter no longer lies with nations, but within the fast-growing cities.

As an example, the adoption by UNESCO of the Recommendation on the Historic Urban Landscape (HUL) in 2011 shows evidence for the ever-growing conflicting issues that arise from the confrontation between heritage conservation and modern urban development. The new HUL approach originated from the concern to preserve not only a monument itself but also its visual perception when integrated into contemporary surroundings (Vienna Memorandum). In fact the HUL approach goes well beyond this principle as "the historic urban landscape is the urban area understood as the result of a historic layering of cultural and natural values and attributes, extending beyond the notion of "historic center" or "ensemble" to include the broader urban context and its geographical setting" (art. 8 of the Recommendation). That the scope of cultural heritage evolved over time from a single iconic monument to a holistic historic urban landscape is indicative of the willingness of conservation authorities to compromise. The UNESCO four pillar paradigm of cultural, economic, social, and environmental objectives (Hangzhou Declaration, 2015) states that the protection and conservation of cultural heritage does not aim to cultural objectives only, but contributes to global sustainable development.

4.3 *Dominant market system*

The fall of the Berlin Wall in 1989, and the subsequent collapse of the USSR in 1991, were historic landmarks in a process that initiated after WWII and that divided the world into two military and politic superpowers, the USSR and the USA. The end of the Cold War eventually lead to the supremacy of the market system, based on economic principles of competition and individual 'freedom to choose', all over the world, including in former socialists regimes.

Although the dominance of the market system globally resulted in better efficiency in allocation of economic resources, and in increasing international trade, it was also accompanied by the privatization of such economic and financial resources, with less public initiatives, and less emphasis put on collective needs and services. This on-going trend may harm the non-profit sector in regions where its management and financing used to rely on public funding and regulations, like the sector of cultural and natural heritage. Today, there are multiple initiatives to develop intermediate models for financing and managing cultural heritage and its conservation. Most are based on Public-Private Partnerships (MacDonald & Cheong, 2014), or on the so-called 'third sphere' of economic resources (neither public, nor private).

Because market dominance impact mostly highly flexible markets (i.e. market with near perfect competition), financial markets were the most involved in new innovative techniques. Research has still to be conducted on how to provide cultural and natural heritage with innovative and tailor-made financial vehicles that could benefit to conservation.

4.4 *Digital revolution*

The impact of the digital revolution on the conservation of monuments and sites cannot be underestimated. The tools of scientific documentation have been multiplied and allow for a total caption of an object or building in its existing state through e.g. 3D-scanning. On the foreground of this revolution is CyArk, founded out of frustration after the loss of scientific knowledge on the Bamiyan statues in Afghanistan destroyed by the Taliban. By using the most advanced technology, 500 key monuments and sites are being digitally preserved for eternity within the upcoming years. While the advantages of the digital revolution for the scientific documentation of cultural heritage are undisputed, there is a lurking danger of inauthentic physical reconstructions when technology is being used to promote nostalgic or ideological illusions.

4.5 *Climate change*

The sustainable development paradigm entails an important environmental pillar, in line with the preservation of cultural and natural heritage. Today's concern with climate change issues is commonly acknowledged as being one of the major threats to cultural heritage. As far as the economic framework is concerned, there is also consensus about the fact that climate change and environmental issues started and developed together with emerging manufacturing industries, increasing economic growth, and globalization. Heritage economics is related to the services provided by heritage properties, as well as to the properties themselves. Facing climate changes issues, the challenge of heritage conservation is to provide sustainable services (like sustainable tourism), sustainable contribution to urban development (urban revitalization), and sustainable production models (local arts, crafts, and skills).

5 A NEW PARADIGM

How can cultural heritage take advantage of creativity and innovation? How can cultural heritage keep pace with social and economic development? How can cultural heritage contribute to sustainable development? How can cultural heritage shape our future? Or will the future determine our relationship with cultural heritage?

Shared values of identity, diversity, and tolerance have been at heart of the mainstreaming of culture and heritage into international policies by the UN, UNESCO, the Council of Europe, or the EU. Simultaneously, technological, economic, social, and cultural changes have always been the motor of innovation within the development of modern architecture, design, and art.

Artistic creativity triggered some process (very similar to the kind of economic changes as identified by Josef Schumpeter in terms of clusters of technological innovations) that goes well beyond the development goals expected to be achieved, giving rise to a new era of profound and durable transformations within society.

Evidence shows that societal changes are the result of cyclical rather than linear developments. Like technological and scientific innovations, cultural innovations by artistic entrepreneurs contribute to Schumpeter's large transitions that contributes to shape and regenerate our society.

The new cultural paradigm is addressing these large transitions that the world is experiencing today. It is embedded in a narrative that goes beyond the contribution of heritage to cultural goals that improve the world. With globalization, growing urbanization, climate change, and economic crisis, heritage conservation is challenged to provide sustainable solutions that improve the world.

One of the main qualities of artists, designers, craftsmen, and architects is that they simultaneously embrace and critically react to technologies, economic, and social developments. Cultural heritage needs to be a part of their sphere of contemporary cultural production, rather than the mere 'heritage' of cultural production from the past. Through a new 'up-stream' paradigm that starts with global challenges and considers cultural heritage as a resource subject to the creativity of contemporary cultural production, 'conservation' will have a better chance of surviving the destructive forces of rapid, merciless urban development. 'Up-stream' refers to the fact that emphasis is put on the contribution of social and economic changes to cultural heritage, rather than vice-versa, such as in the main narratives on cultural heritage today, based on the search for economic arguments to convince policy makers to invest in the conservation of cultural heritage.

By liberating cultural heritage from its presumed and a priory necessity to be conserved, it might actually have better chances of being integrated into contemporary society and in the end be 'conserved' in a more sustainable way.

Applying the 'reverse' logic of this new paradigm within re-use narratives, means adapting historic buildings in such a way that a variety of future functions would be compatible with their main heritage values. This requires a change of culture within the conservation realm where use values are taken seriously and even "forceful re-appropriations" such as the re-use of roman theatres during the Middle Ages (see above) are not seen as a threat but as an opportunity for a second, third, … life for historic buildings. As long as the principle of reversibility is respected along this process, the dangers of this new paradigm can be limited in impact.

The architectural approach to avoid custom-made in favor of flexible projects can also be applied within contemporary architecture, by following the advice of former Flemish *bouwmeester* Bob Van Reeth to build sustainable "intelligent ruins" where the interior structure easily allows adaptation in the future. In Amsterdam the name of "solids" has been given to buildings constructed without a predetermined function and where even the first user of the building can adapt the layout to its current needs (Mensink, 2017).

The program of a building will always be the economic motor of individual projects and cannot come second. Program is here understood not only as the short-term use, but as the variety of functions that a building will receive over a longer period of time.

There are examples of successfully repurposed historic buildings but their result is almost always due to a large amount of time and money, which are not continuously at hand for our growing amount of cultural heritage today. The rapidly developing world, especially in the East, does not allow for a slow reflection period to carefully consider what new function could be custom-made for a historic building.

Therefore a new set of tools needs to be developed, based on the paradigm shift from building to program, from site to the urban social and economic reality. It is by understanding contemporary society and its programmatic needs that the financial means for sustainable development with integration of cultural heritage can be unlocked. Instead of making a building or historic urban landscape the starting point of a long process, this process needs to be reverted. By starting with the programmatic needs of a city and its institutions and organizations, one could start looking for appropriate buildings to fulfill these needs. Cultural heritage will benefit from this reversed logic in the long-term due to a building culture specialized in flexible spatial solutions requiring less structural interventions on cultural heritage.

6 CONCLUSION

Cultural heritage, cultural and creative industries, sustainable cultural tourism, and cultural infrastructure can provide services and welfare to all communities across the world. Cultural heritage and culture-led development generate resilience, innovation, creativity, entrepreneurship, and inclusiveness for individuals and communities through the use of local resources, skills, and knowledge.

By influencing lifestyles, individual or social behavior, and consumption patterns, creativity and innovation contribute to Schumpeter's wave of changes, where cultural heritage plays an important role in the transformation of society, in facing current world challenges and in presenting sustainable solutions. Dynamics for social and economic changes follow a process identified by Schumpeter by 'creative destruction', i.e. the progressive replacement of obsolete structures and organizations by new ones.

Adaptive re-use is an excellent example of such a transitional model. Historic buildings are "transitional cultural objects" that not only help us to understand the past, but that help us understand the future, and without which we cannot even imagine the future. As a discipline, 'conservation' is therefore at the origin of culture itself. (Otero-Pailos, 2014)

The consequences of the current emphasis on cultural heritage by international organizations need to be carefully monitored to avoid a disruption of the balance between keeping what is deemed valuable and upgrading our living environment through the necessary process of creative destruction. This does not mean a destruction of cultural heritage, on the contrary. It means allowing creativity and innovation to foster the discipline of conservation and making sure that cultural heritage can be integrated in a meaningful way in our society today. Faster and at a lower cost.

Further pathways could include research aimed to identify cyclical timeframes in cultural heritage uses, and how such timeframes could be connected to waves of creativity and innovation; the economic crisis acting as a catalyst for changes.

The new paradigm does not go against but follows the United Nations 2030 Agenda for Sustainable Urban Development, the UNESCO Cultural Urban Future Agenda, and the European Union cultural narrative, which aim to achieve international cooperation and economic sustainability.

REFERENCES

Baumol, W.J. & Oates, W.E. (1988). *The Theory of Environmental Policy*. Cambridge University Press.

Broadbent, K.A. (2011). *Assessing the Impact of Local Historic District Designation on Mortgage Foreclosures Rates: The Case of Philadelphia.* Thesis in Historic Preservation, Scholarly Commons. University of Pennsylvania.

Bruno, A. (1996). Fort Vauban, Nîmes. In Mastropietro, M. (ed), *Restoration and Beyond. Architecture from Conservation to Conversion: Projects and Works by Andrea Bruno (1960–1995)*: 61–63. Milan: Edizioni Lybra Immagine.

Carpentier, F. (2014). *Between Architectural Shrine and Urban Resource. Rethinking the Relationship between Heritage Values and Program Renewal in Conservation Narratives*. Master Thesis (promoters Lievevrouw, P. & Bruno, A.). KU Leuven: Raymond Lemaire International Centre for Conservation.

Georganas, I. (2013). The Effects of the Economic Crisis on Greek Heritage: A View from the Private Cultural Sector. *Journal of Eastern Mediterranean Archaeology and Heritage Studies* 1(3): 242–245.

Getty Conservation Institute. (1999). *Economics and Heritage Conservation*. Los Angeles: Getty Conservation Institute.

Globus et Locus. *What is glocalism?*. http://www.globusetlocuseng.org/About_Us/Glocalism/What_Is_Glocalism.kl (Accessed on 2 July 2017).

Knoop, T.A. (2004). *Recessions and Depressions: Understanding Business Cycle*, Praeger.

Licciardi, G. & Amirtahmasebi, R. (eds) (2012). *The Economics of Uniqueness: Investing in Historic City Cores and Cultural Heritage Assets for Sustainable Development*. Washington DC: World Bank.

MacDonald, S. & Cheong, C. (2014). The Role of Public-Private Partnerships and the Third Sector in Conserving Heritage Buildings. *Sites and Historic Urban Areas*. Los Angeles: The Getty Conservation Institute.

Mensink, J. (2013). Solids: radicale innovatie, In *ArchiNed*: https://www.archined.nl/2013/09/solids-radicale-innovatie (Accessed on 11 July 2017).

Ost, Ch. (2017). Strategic Analysis for Cultural Heritage in a Context of Economic Changes. In *Changes: Cultural Heritage Activities, New Goals for Economy and Society*. Key-note Conference, Thematic Week, February 6–8, KU Leuven: Raymond Lemaire International Centre for Conservation.

Otero-Pailos, J. (2014). Transitional Cultural Objects. *Conservation as Innovative Practice*. Lecture 15 May 2014. Leuven: Museum M.

Riksantikvaren (Norwegian Directorate for Cultural Heritage) (2009). The Cultural Heritage and the Economic Crisis Package. The Case of Norway. *Heritage in a Time of Financial Crisis*. Paper Presented May 28–29 at the EHHF 4th meeting. Vienna-Bratislava.

Schumpeter, J. (1934). *The Theory of Economic Development: An Inquiry Into Profits, Capital, Credit, Interest, and the Business Cycle*: 46. Harvard Economic Studies.

Stock, J.H. & Watson, M.W. (1993). Business Cycle. *Indicators and Forecasting*. Cambridge: The National Bureau of Economic Research.

The United Nations. (2012). *Realizing the Future We Want, Report to the Secretary-General, United System task Team on the Post-2015 UN Development Agenda*.

Throsby, D. (2001). *Economics and Culture*: 25. Cambridge University Press.

UNESCO. (2016). Culture Urban Future, Global Report on Culture for Sustainable Urban Development.

World Commission on Culture and Development (WCCD) (1995). *Our Creative Diversity*. Paris: UNESCO.

Innovative Built Heritage Models – Van Balen & Vandesande (Eds)
© 2018 Taylor & Francis Group, London, ISBN 978-1-138-49861-7

The Spinning mill in Sulbiate: A place for changes

R. Moioli
ABC Department, Politecnico di Milano, Italy

ABSTRACT: The paper refers to an example of both an innovative built heritage conservation project and a management model, thus it could be considered as an example of change. It is an ongoing project and it is interesting to highlight how the conservation project and the management issue have been thought to be part of a unique process. The project has been implemented in the framework of Monza and Brianza Cultural District and it is included in the European project CHANGES.

The paper describes the process and proposes reflections on achieved results; the aims are to pinpoint criticalities and strengthens and to create an evaluation system for the management activities.

The paper is mainly focused on the model applied for the formulation of the design process. The case study is interesting as it has been possible to involve the actors in the evaluation process, creating a learning laboratory.

1 INTRODUCTION

This paper presents a practice led research, the conservation process of the Spinning Mill (Ex Filanda ©) in Sulbiate, which is one of the projects carried out in the framework of the Monza and Brianza Cultural District (MBCD), a wide area plan designed thanks to the call for grant launched by Cariplo Foundation.

The framework project will be described first, focusing on the rationale and the expected impacts, but also introducing some figures about the financial aspects, which made more effective the involvement of public subjects and private stakeholders.

The main purpose of the paper are: to describe on the real local chain of decision making, actions, effects, new problems arising, and the willingness and capability of the involved players to relaunch the project; to present a best practice in designing a long term conservation project; to introduce a model of comparison of costs and revenues for the different stakeholders, which has been required to clarify the roles and the perspectives.

The aim is to contribute real experience from the field to the debate on the synthesis of preservation policies and new uses related to Creativity and new entrepreneurship.

2 CULTURAL DISTRICT PROJECT

In this section a short excursus of the general framework and of the long working period will be presented.

In 2001–2005 the pilot project "Magistri Comacini" (Pedrazzini 2005; Della Torre 2005), based on a formal agreement signed by Lombardy Region, Cariplo Foundation and several public and private partners, has been defined and funded with a grant of 6 ml euros. After this first empirical experience, the need for a theoretical framework and a methodology appeared. In 2005 Cariplo Foundation asked prof. Stefano Della Torre to draw up a Pre-feasibility Study in order to define the contents of the call "Cultural Districts (CD) as a tool for economic and social valorisation of local cultural heritage".

In the study the theorethical framework about Cultural District, in the national and international panorama, have been analysed. Florida's, Santagata's and Sacco's models became the bases for the definition of a new type of Cultural District.

The call was launched in 2007 and received 37 proposals. Only 11 have been selected for the first step of the call: the drawing up of the feasibility study in a period of two years. In this phase the total investment for each application was € 85.000, € 60.000 funded by Cariplo Foundation.

In 2010 Cariplo Foundation selected 6 proposals for the implementation of the projects, "each cultural district approved as a grantee after viability analysis by the Foundation was kicked off independently of the Foundation's support. € 20 million is the total amount allocated by the Foundation for its multi-year financial support to the approved cultural districts" (Cariplo Foundation website.)

The six districts are: Cultural District of Valle Camonica, "Dominus" Cultural District, "Regge dei Gonzaga" Cultural District, Monza and Brianza Cultural District, Cremona Cultural District and Valtellina Cultural District.

The contents, the general framework and the rules for the attaining of the grant have been explained in a book (Barbetta, Cammelli & Della Torre 2013); as well as the monitoring and evaluation system, foreseen since the beginning of the process (Della Torre & Moioli 2012).

In 2011 the start up period initiated, the project ended in 2015, and now each CD is following an autonomous path pursuing new activities and trying to evaluate the results.

What is important to highlight is the methodological approach required for the definition of the strategic lines focused on the Built Cultural Heritage (BCH). The challenge was to introduce a shift from restoration to planned conservation in order to pinpoint the economic outcomes, as the new paradigm entails scale economies and cost reduction, and, above all, it moves investments to more qualified activities, such as survey, monitoring, diagnostics, data filing, information management, research, communication… (Della Torre 2010). The new paradigm entails shared methodologies and procedures orientated to a long run vision, to costs control, and has quality as the main goal. The objective is to use the resources for the maximum of positive outputs in a local development process, referring to theories that take into account endogenous factors such as innovation, scale economies, and learning processes (Capello 2011). It is a model of integrated local development in which culture plays a strategic role (Vandesande et al. 2014).

3 CULTURAL DISTRICT OUTCOMES

Before examining in detail the case study, some figures about total expenditures and impacts for the six CD, and a presentation of the adopted methodology have to be presented. It is important to stress that CD project wanted to be an opportunity for innovation and change. Referring to the BCH strategic lines it is worth noting that a relevant change of approach occurred: there was a shift from restoration to preventive and planned conservation.

The new paradigm entails shared methodologies and procedures orientated to a long run vision, to costs control, and has quality as the main goal. In few words, the objective is to use given resources in the way that yields the maximum of positive outputs in a local development process.

The first factor of change is related to the conditions for obtaining the grant: with the previous rules there were no controls about procedures and quality of the works, no evaluation of the quality of expenditure and of the whole process in terms of impacts, and any monitoring system was foreseen.

The monitoring and evaluation system is a web based system for projects and strategies, based on the collection of data on the progress of single strategic actions. This data should allow to check whether:

- foreseen activities are actually carried out;
- intervention on tangible heritage is achieved;
- different foreseen procedures are followed for each sort of intervention;
- schedules are respected;
- how, how much and by whom works carried out under grant funding are used.

The evaluation system works out the information filed for monitoring every semester, no longer by single actions but by two macro-categories, that is for the evaluation of the process in the actual context of the District, and for the evaluation of the impacts in the District areas.

Cariplo Foundation followed the four years (on average) start-up period of the six CD and at the end of such lapse of time it has been possible to rework all the data coming from the monitoring system. The total expenditure for the six CD has been of € 58 292 513,00 (Table 1) and it has been sustained for 32,86% by Cariplo Foundation and for 67,14% by the local subjects, both public and private (Table 1).

From these data it is possible to obtain percentages which clarify the weight of the different sources of funds, the different typologies of expenditures and the impact of conservation activities on the total amount.

The expenditure for the conservation of BCH has been analysed more in detail: it is the 75,30% of the total and it is important to highlight the fact

Table 1. Description of the expenditures.

	€	Number
Total expenditure	58.292.513	
Cariplo Foundation Grant	19.152.968	
Cultural Districts expenditure	39.139.4545	
Total expenditure for Conservation	43.895.193	
Total expenditure for other activities	14.397.320	
Cariplo Foundation grant for Conservation	12.585.399	
C.D. Expenditure or Conservation	31.309.794	
Total actions		123
Conservation actions		45

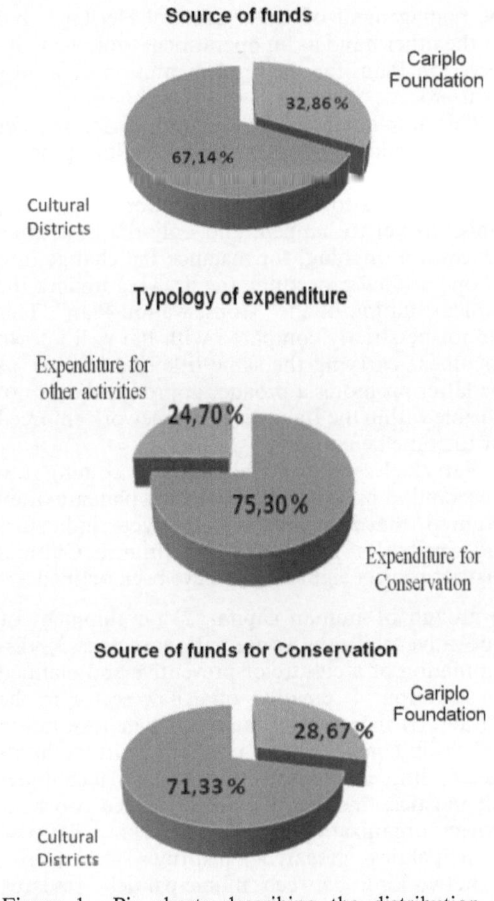

Figure 1. Pie charts describing the distribution of expenditures.

that in this case the amount of the grant for BCH is lower—in percentage—than the co-financing of the other activities (Figure 1).

Cariplo Foundation's aim was to highlight the active role of Cultural Heritage within local (economic) development strategies. A deep analysis of data is ongoing and they are under process according to a set of indicators aimed at evaluating the impacts on private sectors.

In 2011 the start up period initiated, the project ended in 2015, and now each CD is following an autonomous path pursuing new activities and trying to evaluate the results. At the moment it is possible to present a synthetic preview of the results. Regarding the impacts on local economy, the analysis of Cariplo Foundation has pointed out that about 700 enterprises have been involved and the majority are small-medium enterprises (SME), according to the definition of European Commission.

As a general summary it is possible to state that the expenditure has been distributed among the following activities: conservation and construction works, professional fees, art, sport and leisure, manufacturing, communication and dissemination.

For the two first categories the expenditure has been more than 10% of the total amount, while for the other categories it varies between 4% and 8%.

The main remark is that conservation activities are the main cost items and the related number of enterprise involved is in inverse proportion. On the other hand for professional fees, communication and manufacturing there are a relative high number of subjects for a low expenditure.

4 MONZA AND BRIANZA CULTURAL DISTRICT

Monza and Brianza Cultural District (MBCD) has been thought as a model for a local development driven by Culture and BCH. Before analyzing the case study it could be useful to present the general framework in order to make clear the methodological approach, the conditions of the grant and the procedures adopted for all the interventions carried out within the MBCD broader project.

The main strategic vision is focused on culture as a unifying factor and on the creation of synergies between tangible and intangible heritage through the implementation of some specific actions such as conservation works, the front office for Preventive and Planned Conservation and the enhancement and development of cultural and creative activities.

All this is possible only by the involvement of different supply chains and stakeholders; in this case the chosen sectors are: public bodies, education sector, category associations, innovative productive sectors.

The MBCD is composed by 12 partners which have signed a formal agreement both on the contents and on the economic contribution. The foreseen total investment is € 8.5 millions, € 3.6 millions funded by Cariplo Foundation.

The aims are the creation of a stable network made of cultural system and economic system and, more specifically, synergies between cultural heritage and creative enterprises. Thus the strategic lines are: Culture; Identity; Training; Innovation.

Several actions have been planned for each one of these strategic lines, trying to create links and common activities. For instance, Culture, Identity and Training shared some work-packages as Heritage Education Activities, courses of Cultural Project Management and Planned and Preventive Conservation. Culture and Identity have been connected by means of the restoration works and valorization activities, based on the knowledge generated by the conservation process, and of the Front Office for Preventive and Planned Conservation. Training and Innovation have in common all the actions, courses and mentorship, for the support to new cultural and creative companies.

The designed actions can be summarized in: conservation, enhancement and fruition of cultural heritage; creation of opportunities for process innovation in the local economic milieu; training.

In total there were 13 actions, seven of which directly connected with the Conservation and Valorisation of BCH: four restoration intervention (Ca' dei Bossi in Biassono, Da Corte Castle in Bellusco, the Stables of Borromeo Arese Palace in Cesano Maderno and the Spinning Mill in Sulbiate), the front office for Preventive and Planned Conservation, training courses and valorisation activities such as pH – Performing Heritage.

The twelve partners have been involved in different actions as one of the goal was to create networks and sharing communities. Actually the most interesting datum is the number of people who participate in the process, for the whole period or only in a specific phase. The total number is superior to the hundred people if we consider both the internal staff of the subject partners and the suppliers chosen—often with an open call—for the single activity, such as valorisation activities. Furthermore, there are all the participants of the training courses and of the several communication events.

Given the high number of involved persons, it is possible to state that they represent different sectors of community: policy makers, stakeholders, citizens and users.

As mentioned above, the tailored set of rules is an important aspect. A second factor of change could be identified in the modified approach toward the conservation process that can be summarized as follow:

- restoration is one phase of the conservation process;
- the restoration design proposes and plans the new function
- the project has to be accompanied by a management plan
- a Conservation Plan for a medium-long term has to be provided;
- the valorization activities have to be designed as well.

In brief the keyword is: management. All the activities entailed in the long term process—protection, conservation, valorisation, communication and public enjoyment—have to be coordinated in a systemic vision. The objective is to provide a methodological framework of Planned and Preventive Conservation, meant on one hand as a strategy for the management of Built Cultural Heritage, and on the other hand as an operational tool, the Conservation Plan, for the implementation of a long term process.

This implies a conceptual transformation for the understanding of the conservation process. In Italy the transition from the idea of a conservation process to the implementation of effective tools has yet to happen, and only the first experiences are ongoing, for instance the change into an operational procedure, the drawing up and the implementation of the "Conservation Plan". This cannot be strictly compared with the well known document carrying the same title (Kerr 2013), as the latter promotes a broader approach that is not feasible within the Italian legal framework enforced for the time being.

Nevertheless managerial culture is coming into preservation practice, as monitoring plan are often required, that is a clear set of targets, indicators and timetables. In Monza and Brianza Cultural District project eight targets have been defined:

1) growth of human capital; 2) development of innovative skills in conservation sector; 3) dissemination of a culture of preventive and planned conservation; 4) creation of a new sector in the market; 5) increase of the quality in restoration and maintenance works on built cultural heritage; 6) implementation of advanced techniques (diagnostics, monitoring, ICT, skilled workers, project organization...); 7) networking between municipalities—research institutes—enterprises; 8) networking between municipalities—research institutes—enterprises.

A change of mentality is needed which can happen only through a cultural action aimed at the creation of a social/intellectual capital that is able to understand cultural heritage in terms of cultural capital. Thus education, the professional training and the qualification of the operators in the conservation sector are essential topics.

The four restoration projects have been designed taking into account the functions and the management model, and it is possible to link each intervention to the strategic lines. The Da Corte Castle and the Stables of Borromeo Arese Palace are the projects for the implementation of Culture and Heritage, Ca' dei Bossi and the Spinning Mill are referred to the strategic line for Innovation and entrepreneurship.

5 THE SPINNING MILL IN SULBIATE

The restoration of the Spinning Mill in Sulbiate is one of the thirteen actions of Monza and Brianza Cultural District.

Figure 2. The Spinning mill interior before the restoration – Photo credit: R. Moioli.

Figure 3. The Spinning mill interior after the restoration – Photo credit: E. Mascheroni.

The Spinning mill is an industrial building dated back to the Twenties of XX century and abandoned since 1970s (Figs. 2, 3 and 4). In 2009 the Municipality signed an agreement with the private owner who transferred the property to the public body as a tax burden for the edification of a new residential complex. The operation is justified by the high symbolic and emotional value attached to the building by the community. This proposal has been selected as one of the four restoration works in the District project and granted mainly for the idea of the new function: a place to create innovation through art, technology and social capital.

In 2010 the first procedural document, the construction briefing or "preliminary document for the intervention design", established all the future procedures and the methodological framework. In the two-years period 2011–12 the first act was the signing of the formal agreement with the Monza and Brianza Province, and this act formalized the beginning of the start-up period.

Figure 4. The Spinning mill exterior after the restoration – Photo credit: E. Mascheroni.

From the operational point of view in these two years the main activities were: the call for bids open to professionals for the design based on "the more convenient proposal" model, that means quality assessment criteria; the application (as political turbulences produced the need for additional financial resources) to the ASTER competition, a call issued by Lombardy Region for entrepreneurial projects; the launch of designing phase.

In 2013 the preliminary and final designs have been delivered and they were the base for the second call for bids, according again with the same public procedure, for the working plan and works contract. From 2014 to 2016 works have been carried out and in September the opening ceremony took place.

At the end of the start-up period it has been possible to begin the assessment both of the process and of the achieved results.

The project has been drawn up with the participation of 14 partners, among them: Monza and Brianza Province, Sulbiate Municipality, Chamber of Commerce, Apa Confartigianato, Politecnico di Milano–DESIGN and ABC depts., AFOL, Green High Tech District Foundation, Enea Agency.

Another point of interest is that ASTER call, promoted by the Regional Direction for Industry, Craftmanship and Agriculture, was for the implementation of new activities, coherent with the local productive vocations, thus not strictly for cultural purposes. But to bring culture and economy to work together, under the flagship of innovation, was exactly the vision of Distretti Culturali project, and this story comes to confirm that partnerships between different stakeholders make the projects more sustainable and resilient. The total amount of the grant was € 895.000,00 for the restoration of the Spinning mill and the implementation of support activities to the start-up companies.

Therefore this story is a good example of value created "upstream" (CHcfE Consortium, 2015), that is during the negotiation phase that prepares the intervention. Only the network created thanks to the CD has been able to design a successful proposal for the ASTER call and to allow the Municipality not to lose the Cariplo Foundation grant, but to match the grant with funds destined to new entrepreneurship, not necessarily to heritage preservation.

The activities carried out during the four year period show many positive results, but also some criticalities. For each achieved result it is also possible to present both positive and negative aspects.

The choice of using public procedures based on quality and not only on the minimum offered price has been a successful option and a best practice, as usually this type of tender is considered a hazard, difficult to put into practice and expensive for the nomination of a committee of experts. Instead, it has been demonstrated that in this way competition has been stimulated: professionals and building contractors have been obliged to understand the needs of the client, to share the methodological approach and to provide well thought-out proposal, also from the economic point of view.

This procedure however takes some additional time, and in this specific case three public tenders have been foreseen, deciding not to have the same architect in the three design steps. This can be considered as a criticality, as the restoration project is firstly a cultural matter and not only a sequence of technical solutions, and the role of the architect is fundamental.

One of the public tender has been drawn up by the Province staff in collaboration with the Municipality officers. It has been a test of collaboration among different public bodies. The positive aspect has been the support provided under the administrative point of view, the main criticality has been the excessive time needed for the bureaucratic steps.

Since the first designing phase, many communication activities have been implemented. A first successful opening of the Spinning mill, still in bad repair conditions, hosted almost five hundred visitors: during the annual event called *Ville Aperte* (open monuments), the Spinning mill has been opened to the public and it became the stage of a site specific performance of contemporary theater and dance, created in the frame of *pH - performing Heritage* event. In the following years the Spinning mill has been regularly opened to the public for the annual *Ville Aperte* event, also during the restoration site.

Besides these communication activities, a more participative action has been put into practice. Thanks to the ASTER funds, it has been possible to involve Politecnico di Milano – Design dept. for the creation of a working group composed by local people and stakeholders, in order to spread the habit to bottom up processes and to train a group of social actors able to apply to public tenders, such as the forthcoming one for the management of the Spinning mill once the restoration would be finished.

The interesting features are the bottom up approach and the attempt to integrate the restoration issues with the future function, taking into account also the management model.

When in 2016 the restoration works finished, the Municipality published the public announcement for the space/activities management. The call required subjects able to:

- think innovative ideas;
- create contamination processes among art, design, making, handicrafts, media and food:
- demonstrate to have the financial structure for the management of the building and of the valorisation activities.

In the call it is clearly expressed that the aims of the Spinning mill project "*La fabbrica del saper fare*" are to generate new production processes and to activate partnership with enterprises, craftsmanship, public bodies by means of mentorship, valorisation of micro-companies for not mass-produced objects. Therefore the call reflected the purpose to enhance built heritage as a driver for actions promising positive and sustainable effects on local community, economy, environment and culture, taking into account time as a decisive factor (Moioli 2015).

The winner of the public competition has been a group of three different subjects (website ExFilanda):

- *Consorzio Comunità Brianza,* a social enterprise based in Monza and Brianza region. Its main goal is to strengthen and to qualify the activity of the associations members of the Consortium. It is a subject created for playing a political, social and economic role promoting projects with and for people;
- *Make in progress*, an association for social purposes. It has been developed within the participative project led by Politecnico di Milano–Design Department. The aim was to create a community based on participatory mechanisms and focused on the topics of making (fab lab), art and food. It is an open community;
- *DelleAli,* a cultural association. Their main activity is theatre performances, but also social projects carried out through theatre experiences. The central role is played by the «person» who is always on the floor.

The period from September 2016 to the first months of 2017 has been spent for communication activities and events, two years planning and definition of the juridical characteristics of the legal subject in charge for the site management.

The Municipality and the subjects in charge for the management of the site defined the details of the contract and of the governance of the project. It meant a precise delineation of the roles of each subject, of the monitoring and of the expected outputs, as the objective is to promote a place for private workers maintaining a public interest.

This was a first attempt of an improvable best practice. Maybe it would have been better to choose the subject in charge of the management before the restoration intervention. But the Italian law for the concession of a public space/building foresees a public tender, which in the technical attached documents should mandatorily include all the information about the characteristics of the building and of the systems, as well as the costs for the ordinary run management of the building. The participants to the call have to use such information for the submission of a detailed economic-financial prospect and a business plan. It is clear that in case of a restoration intervention, it is not so easy to comply with these rules. Probably it would be necessary to create an even more tailored process.

At the end of this path, the formal agreement has been signed and the most relevant conditions are: the building keepers pay a rent to the Municipality and undertake the ordinary operating costs; furthermore, 2 open days per year and 4 events per year involving local schools must be guaranteed; the typology of activities have to be submitted to a scientific committee and training courses and networking with local social/economic subjects have to be included in the annual programme.

The Spinning mill, in a space of 808,68 square metres, hosts a maker space, a co-working area, a centre for artistic and cultural experimentations, a food and drink corner. This space would like to become an attractive place for local communities.

6 THE ASSESSMENT

After the closing of the restoration site, the evaluation activities, foreseen in the JPI project CHANGES, started with the collection of all the data coming from the monitoring system filled in during the start-up phase and of all the design and site documents included the Maintenance Plan.

But the most interesting information came from the periodic interviews to the working group composed by representatives of the three subjects in charge for the management. In other words, the attempt has been to carry out a theory-based evaluation (Rogers 2008), through the participation of players in discussions, in which their rising awareness is the most important evaluation criteria.

The aim of this activity is to assess both the impacts on local (regional) milieu in terms of creation of innovative processes and the sustainability of the challenging path undertaken by the small Municipality of Sulbiate. The Spinning mill was a private building, probably destined to the demolition, which was acknowledged as cultural heritage for its intangible value linked to the collective memories of the inhabitants. The challenge has been to acquire and restore it with the less expenditure as possible, to avoid to transform it in an economic burden not sustainable by the public body, to return it to the people as a cultural and public good, but with a social and economic role. Thus the evaluation model has to take into account the quality of the process, detecting strength and weak spots, the achieved results from a quantitative point of view and all the cost-revenues indicators.

The result of the periodic meetings prompted the sentiment expressed above while telling this long story: that is enduring discussions on the hardly solved problems, but also awareness of the developing process, the targets, the recognition gained in the region, where the Filanda is becoming a model.

Many issues still concern financial sustainability and the mutual duties: that's why a model has been structured using criteria and indicators as shown below. The model is quite simple, but useful as it encompasses many items, which players were not ready to consider, thus going into the risk of definitely mistaken evaluations. As the process has just started, data are not yet available for the first balance, but the costs structure can be shown.

A first section includes the costs, starting from the conservation activities from the beginning of the process. The amount of money spent for the restoration works can be partitioned in the following cost items:

- Cultural District staff;
- administrative activities;
- works;
- communication/valorisation activities.

Then there are the expenditures, in the normal operation phase, for the management of the Spinning mill. They are related to the conservation activities on the long term, to use and valorisation. It means that rent, spending for the bills/cleanings/facilities, local taxes and costs for implementation of the activities must be taken into account. It is important to consider not only the amount of the expenditure, but also the typology of funders, as some remarks can be made depending on the distribution in percentage between public and private subjects.

According to this criterion it is interesting to compare in detail the costs suffered by the public owner and the private subject responsible for the management of the building. The Municipality is in charge for the costs (if any) of the scientific committee, the implementation for the first 4 years of the Preventive and Planned Conservation activities and the related administrative procedures. The costs sustained by the keeper are: rent, insurances, bills, safety, management costs of each organizations (taxes, salaries, specific insurances, etc.), and communication activities.

As costs, revenues as well must be calculated, following the same logic. For the public owner the rent is an income, speaking from the financial point of view, but it is important to assign a monetary value also the public fruition guaranteed by the use of the spaces, the 50 days per year available for public initiatives and the valorisation activities for free offered by the keeper. The private subject takes profit from the following activities: core activities of each one of the three organizations, sublet of spaces, training courses, production and sales of design objects, drink and food and sponsorships. Furthermore, it is worth notice that the nature of the project make this subject eligible for the obtainment of public funding.

7 CONCLUSIONS

This project has the ambition of demonstrate that BCH can be a resource and not only a cost for the public budget and for collectivity. Furthermore there are other points of interest: the effectiveness of a well planned conservation process; the different roles of public and private subjects in a partnership focused on cultural, social and economic activities; the community – or communities – involved in the past but also in the future of the Spinning mill. Regarding the first topic, the evaluation model will provide an answer. While for the other issues some final reflection can be made.

The evaluation of the impacts generated by the restoration has to consider the whole process: the framework project (Lombardy Cultural Districts), Monza and Brianza Cultural District and then the Conservation process of the Spinning mill triggered in a long run vision.

The Municipality had always a clear vision of its tasks and its role: a public body has the responsibility of making the public investment effective; the objective is to guarantee the collective interests; the duty to preserve the building. In a partnership with a private subject its role is to control and to support.

At this point of the path some questions still remain open:

- Which should be the breakeven point between the public funding and the private investment?
- Private subjects should be financed again by public sector (as they are «working» in and for cultural heritage)?
- How the conservation of the building can continue to be a priority?

The case study made many data and information available, but above all it made possible to involve the actors in the evaluation process, creating a learning laboratory perhaps able to answer to the questions mentioned above.

REFERENCES

Arquilla, V. & Barbieri, A. (2015). Design as enabler of anticipatory systems: the MakeinProgress case study. In *1st International Conference on anticipation*. Trento 5–7 November 2015.

Arquilla, V. & Zinzone, M. (2017). Design e Making per lo sviluppo di lavoro, apprendimento e socializzazione. L'Ex Filanda di Sulbiate. In Bisson, M. (ed), *Environmental Design*: 133–146. Milano: De Lettera.

Barbetta, G., Cammelli, M. & Della Torre, S. (eds) (2013). *Distretti culturali: dalla teoria alla pratica*. Bologna: Il Mulino.

Capello, R. (2011). Location, Regional Growth and Local Development Theories. *AESTIMUM*, 58: 1–25.

CHCfE Consortium. 2015. *Cultural Heritage counts for Europe*. Krakow: International Cultural Centre.

Della Torre, S. (2005). L'integrazione dei sistemi culturali come strumento di tutela pro-attiva. *Arkos*, 10: 20–25.

Della Torre, S. (2010). Economics of Planned Conservation. In Mälkki, M. & Schmidt-Thomé, K. (eds.), *Integrating Aims. Built Heritage in Social and Economic Development*: 141–156. Helsinki: University of Technology, Centre for Urban and Regional Studies Publications.

Della Torre, S. (2015). Shaping Tools for Built Heritage Conservation: from Architectural Design to Program and Management. Learning from 'Distretti culturali', in K. Van Balen & A. Vandesande (eds), *Community Involvement in Heritage*: pp. 93–102. Antwerp – Apeldoorn: Garant.

Della Torre, S. & Moioli, R. (2012). Designing an active monitoring system: the Planned Conservation Project in Monza and Brianza Province. In S. Mendes Zancheti & K. Similä (eds.). *Measuring Heritage Conservation Performance*: 142–147. Olinda & Rome: CECI & ICCROM.

Florida, R., Lee, S.Y. (2001). Innovation, Human Capital and Diversity, Carnegie Mellon University.

Kerr, J.S. 2013. *The Conservation Plan: A Guide to the Preparation of Conservation Plans for Places of European Cultural Significance*. ICOMOS Australia.

Moioli, R. (2015). Architectural Cultural Heritage and Sustainability: How Many Pillars? BIOCultural 2015. International Conference on Sustainability in Architectural Cultural Heritage, Pafos pp. 201–211.

Pedrazzini, L. (2005). Patrimonio culturale come volano per la crescita: l'AQST "Tremezzina-Magistri Comacini", *Territorio*, 32.

Rogers, P.J. (2008). Using Programme Theory to Evaluate Complicated and Complex Aspects of Intervention. *Evaluation*, 14, 1: 29–48.

Sacco, P.L. (2003). Il distretto culturale: un nuovo modello di sviluppo locale? *Ottavo rapporto sulle fondazioni bancarie*, year LI, n. 3.

Santagata, W. (2005). Cultural District and Economic development, EBLA Center, Università di Torino.

Therond, D. & Trigona, A. (eds) (2009). *Heritage and Beyond*. Brussels: Council of Europe Publishing.

Valentino, P.A. (2003). Le trame del territorio. Politiche di sviluppo dei sistemi territoriali e distretti culturali, Milano.

Vandesande, A., Van Balen, K., Moioli, R. (2014). Costing the built environment: towards a policy of strong sustainable development. *XXX Convegno Internazionale Scienza e Beni Culturali "Quale sostenibilità per il restauro?* Bressanone, Arcadia ricerche, pp. 457–467.

Websites

http://www.fondazionecariplo.it/en/projects/arts-and-culture/distretti-culturali/i-6-distretti-avviati.html

https://www.youtube.com/channel/UCznz_S9owlaCTUIFI3TempA

http://www.distretticulturali.it/

http://www.distretticulturali.it/monza-e-brianza/

http://www.distrettoculturale.mb.it/distretti_culturali/attivita/villeaperte/pH/

Innovative Built Heritage Models – Van Balen & Vandesande (Eds)
© 2018 Taylor & Francis Group, London, ISBN 978-1-138-49861-7

Building conservation as a catalyst for regional sustainable development. Halland model as planning and acting sphere

C. Gustafsson & A. Ijla
Uppsala University, Sweden

ABSTRACT: Recently, conservation of built cultural heritage has been understood as a driver for sustainable growth and is no longer just recognized as an obstacle to development. Sustainable integrated conservation is regarded as an important strategic investment for the entirely society and as a catalyst for sustainable development. Halland Model states that utilizing "trading zone" as a democratic pluralistic arena in historic conservation's decision-making process is the catalyst for sustainable regional development. The model developed where the "trading zone" defined as an active melting pot aiming at assisting various actors for making sense together. This paper proposes changes and enhancement for the model in order to achieve an integrated vision of historic preservation within regional development planning process. In this way, collective governance sphere "the enhanced model" is constitute by the stakeholders and the members of an inter-communicating community, situated in the particularities of time, place, and regulations aiming at guiding and managing sustainable long-term actions.

1 INTRODUCTION

The Halland Model was a regional strategic cooperation between the construction industry, regional government bodies and the cultural heritage sector (Gustafsson, 2011) aiming at sustainable development. In times of recession, the objective was to increase the volume on the construction market which would work as an accelerator to the whole regional economy. Focus was on investments in conservation projects and new business on the improved premises. Cultural heritage hereby got the role as catalyst for regional development.

Nowadays, all the regions in the European Union are operating regional development programs, aiming at sustainable development, with new solutions for innovative strategies strengthening regional competitiveness through smart specialisation. This is a new challenge for the cultural heritage sector, which often has been targeting to the city planning development as well as the policies and strategies level of sustainable regional development. Here, conservation officers and part of the government representatives who are engaged in development process of built cultural heritage have often been given the role to say "no" to changes, while initiatives to make changes have been given to developers and private sector etc. The Halland model and trading zone give the opportunity for various stakeholders with different objectives and interest to present their objectives, strategies, resources, values and goods to achieve the established goal of regional development. The Halland model managed to bring different actors to share and to trade their values and interest through face-to-face negotiation, bargaining in one collective forum to engage in consensus-oriented decision making. Despite that trust building and shared understanding shows that these forums focus on "small-interim wins", the model needs to be development to face the shortage of financing and deficiency of successful public policy. Collaborative governance gives a momentum enhancement from taking decision together to working and acting together aiming to ensure the continuity of historic conservation as strategic engine for sustainable regional development.

2 HALLAND MODEL: AN OVERVIEW

The Halland Model for sustainable building conservation is a theoretical and practical-experienced model aiming at nurturing regional development based on sustainable principles. The partners were the construction industries (contractors and trade unions), regional government bodies (County Administrative Board and County Labour Market Board) and the Regional Museums of Halland. The partners had clear and defined areas of responsibility; for instance, the Labour Market Board funded the projects to greater extent and also made sure to avoid displacement effects, the contractors were responsible for construction guarantees, the trade unions to select construction workers, and the cultural heritage sector for conservation qualities. All together the partnership developed methodologies for selections of buildings as well as new activities in the conserved houses.

The regional joint venture includes job creation, workforce development, the construction industry, the historic built environment sector, real estate owners, local and regional authorities and trade unions. The conservation projects (the footing and grounds of the Halland model) were started during 1990s recession aiming at increasing the total volume of construction, prevent bottle-neck problems and to absorb the unemployed construction workers within the projects' activities. The main objective of the scheme was: Save the jobs, save the craftsmanship, save the buildings. A fourth motto was introduced already in the first year: to establish new activities aiming at inclusive, sustainable and innovation-driven development on the improved premises. The model was soon developed into a regional cross-sectoral joint-action network aiming at sustainable growth. The aim also was strengthening competitiveness, sustainability including use of renewables and reuse/recycling of materials, adaptive re-use of the conserved and restored historic buildings, and development of planning building conservation initiatives and investments.

Christer Gustafsson's (2011) initiated the Halland Model as an application oriented theoretical platform and a new historic conservation model, providing adequate approaches to solve the economic challenges in a sustainable approach. A basic and entrepreneurial model is developed where the "trading zone" is defined as an active arena for negotiations and exchange of services or a field of force corresponding to the actors' policies, values, facts and resources.

In the Halland Model, more than 1,200 building construction workers and apprentices were employed and then trained in traditional building techniques operating in conservation projects at about 100 historic buildings at risk, under supervision of skilled craftsmen and conservation officers. Selected objects include castles, windmills, industrial sites, dwellings, warehouses, theatres among others.

After the completion of conservation work the improved premises made new functions available inspiring the start of several new businesses. The chosen buildings were seen as resources to be taken advantage of and to develop. Preserved built environments are often seen as attractive for dwelling and if used properly they can be an integral part in trade activities and other businesses and even increase their market value. This is one of many added values which have come out of the model. About 1,500 contractors and suppliers have been involved and about 400 new jobs have been created directly depending on the execution of the Halland Model and about 200 indirectly. The Halland Model has been exported e.g. to Russia, Poland, Lithuania and Iceland as regional project for sustainable development and the experiences of the Halland Model has been disseminated in several conferences all over the world.

The overarching Halland Model gives good examples on managing of environmental and economic performance without diminishing the cultural value and social history in built heritage (Ferilli, Gustafsson and Sacco, 2017). Adaptive reuse of historic places, spaces, and spatial implies that the concept is not only building on the three pillars of sustainability i.e. environmental, social and economic, but also manages to enhance the five capitals: human, physical, natural, social, and symbolic/cultural capital. The Halland Model has a direct and indirect effect and multipliers within local community and regional levels.

3 METHODOLOGY

The region of Halland has been applied as the research methodology due to the fact that it was identified as heritage conservation area based on Halland Regional Museums which reported on approximately forty historic buildings at risk, to be suitable for conservation in the framework of the Swedish government stimulation and fund package aiming at revitalizing labour market sector within the construction industry. Robert Yin (2014) defines case study research method as an experiment which investigates a contemporary phenomenon within its real-life context. In this project, the researcher seeks to answer of the hypothesis by selecting small geographical area as the subject and investigate contemporary real-life phenomenon through analysis of a limited number of events. The information gained mostly comprised of opinion based on facts, personal experiences with Halland model project, supporting data, information and objective evidences.

Besides that, in gaining primary data the researcher also conducted constructed interviews and goal-oriented interview among community members, workers, buildings owners, local labour market board, municipal authorities, construction sector, heritage sector etc. From the personal observations and experience of the author (Gustafsson, 2011) and interviews, the researchers know how reliable the data is and that it is up-to-date. Understandings gained about different concepts and issues of heritage conservation from secondary data were accumulated and compared with the findings from the author's observations and stakeholders' interviews. Study on implementation and application of conservation plan by municipal authorities and the project leader were also conducted. From the case study and literature review, an evaluation of the current strategies and mechanism of the Halland model used by selected municipal representatives was done.

Through the author observations conducted on designated buildings and sites, evidences of cultural, social, and economic activities can be addressed. Halland building conservation model has been targeted cultural, social, and economic buildings with significant historical values within the area of study. Throughout the observation and interviews it was found out that these buildings have an architectural significance that makes the character of Halland's historic image.

4 RESEARCH QUESTIONS

Peter Galison introduced the concept of trading zone as a useful instrument for understanding innovation processes in the field of science (Galison, 1997). This paper expands upon previous work also by Alessandro Balducci and Raine Mäntysalo (2013), Christer Gustafsson (2011), Salvador Muñoz Viñas (2005), and Sverker Sörlin (2001) regarding to the trading zone, and it evaluates the sustainable effects Halland model and historic conservation projects on boundary-spanning challenges for regional development, strengthening competitive and innovative advantages of the region. The research design allows the paper to address the following research questions:

1. Does built cultural heritage has a positive regional development impact as a result of the special value created by building conservation projects?
2. Can "the trading zone" as an active decision-making sphere be compatible with the rise of the economic and political worries in Europe?
3. How can collective decision-making process be enhanced into to present economic and political environment for knowledgeable and collective "deciding and acting"?

5 HALLAND MODEL: FROM COLLECTIVE DECISION MAKING TO COLLECTIVE GOVERNANCE

A conflict among different stakeholders on the preservation of historic urban fabric, place, and site from one side and development changes on the preserved character has formed the central discussion for conservation in meaning and practices. The conflict has expanded from the dispute on heritage and conservation values to become an argument about decision making, management, and planning of historic place or area. This article evaluates Halland Model and the trading zone notion in terms of studying the current conflict among the stakeholders and argues for a sustainable governance approach to manage and planning of heritage decision-making process based on communicative, collaborative and cooperative actions together within Halland and Trading Zones models.

To those who are seeking practical solutions to building conservations problems, Halland Model and the Trading Zone may offer the theoretical answer. However, there are operational procedures which might be invited. Developing skills in translation, in constructive approach, in collective and understanding actions to be able to realize the potential of strategies understood as cooperative-collective discussion, addressing, and working out and to prepare the stakeholders to act in respect of common concerns about the sustainable development role of historic conservation in a certain region.

Development of Halland Model needs to change the mind-sets and practices evolved within the trading zone arena to identify the stakeholders' potential within a new communicative, dialogue-based, and cooperative governance system. Long life skills are so needed in order to take into the collective thinking in respect to the cultural heritage conservation's public concerns in a creative process. In a contemporary knowledge-based economy and increase of various multidisciplinary-dynamics networks, through developing a cooperative, communicative, and collaborative governance process such Halland Model or Trading Zone notion, it is possible to develop a future-progressive force not only to make sense together, but also to govern together.

Collaborative, communicative, and cooperative governance process of Halland Model or Trading Zone notion have emerged as a response to the failure of decision-making process and the lack of consensus on the importance role of cultural heritage conservation among the engaged parties as well as due to high cost of politicization of rules and regulations. The model argues that the focus is on decision-making in the cooperative, communicative, and collaborative manner where groups of specialists, organizations, systems of public institutions, private sector, and common making decisions. The actors work collectively in distinctive ways, using particular processes, skills, and capacity for the provision of public goods.

Despite Halland Model and trading zone notion propose collaborative forums aiming at reaching a consensus-oriented decision making and the models focusing on public policies and issues, the models managed to drive the actors to engage in "positive-sum bargaining" and succeeded to encourage them to develop cooperative short-term alliances. Thus, the result of both models lead to ad hoc communication and cooperation and they failed to transfer conflicts into long-run cooperative governance.

In this paper, the main intention to utilize the SWOT analysis of Halland Model in order to treat the weak points and obstacles to understand how it is possible enhance the governance part of the model. Although scholars in policies and strategies who are studying governance systems had already made very important theoretical notions, however, no one studied or standardized a governance theoretical collaborative, cooperative, and communicative approach of heritage sustainable development.

This study investigates the impacts of Halland model over the past 25 years and used the SWOT analysis to develop Halland model, not only as a sound idea, but how to fill the gap of good governance cooperative and collaborative act together. The model enhancements are derived by the earlier experience and the evaluation of the model outcomes. In this study, focus is to understand the conditions under which the stakeholders and actors cooperate within the trading zone. How they achieve consensus and agreement? Which process and in which negotiations' atmosphere? By finding answers for our interview questions, we discovered that trading zone interested and satisfied with the outcomes rather than policy and process outcomes.

Central enhancement steps have been made to elaborate Halland model. The model has three broad improvement components: developing capabilities and capacity; planning, building vision and strategy; and win-win policy making. Each of these broad components can be disjointed into more detailed components and tasks. Trading zone and The Halland Model have treated as the core of the enhanced governance model, while these three components represented as major contributions for the innovative-collaborative process. The following describes the three components in more detail which form the innovative-collaborative process of Halland model.

5.1 *Developing capabilities and capacity*

This part of the model enhancement aims at facilitating and encouraging the cooperation among stakeholders. Breaking the division and removing the pre-history conflict helps the stakeholders to have a shared vision for what they would like to accomplish through collaboration on mutual respect basis. The ambition of this phase is to create power balance between stakeholders within the trading zone hub. The balance among stakeholders can be created through building the capacity of the fragile partners to have an equal footing with other stakeholders. The process aims at constructing a process by strong actors.

Additionally, leadership considers as a critical component in bringing parties together and to lead a cooperative-collaborative spirit and process. Successful mediator leads to constructive negotiations. Leadership can set and maintain clear ground rules, building trust, facilitating dialogue, and exploring mutual understanding. Lasker, Weiss and Miller (2001) argue that collaborative leaders must be skilled in terms of promoting an active participation, ensure sustainable and control over the dialogue, and facilitate productive group dynamics.

Institutional or organizational building refers to building the capacity of those fragile stakeholders. Empowering weaker representatives helps to bring a balance of power among stakeholders and gives meaningful and positive participation. Institutional building also refers to formulate accurate and constructive protocols, ground rules for cooperative and collaborative process. Trading zone must be inclusive for all potential partners and this leads to a process where policy outcomes and decisions represents a broad-based consensus. Non-inclusive representation threatens to undermine the legitimacy of policy outcomes and doubts the decisions taken by the trading zone forum.

On the other hand, trust building among the stakeholders is important aspect of the early collaborative and cooperative process. Dialogue and negotiation, and bargaining can be quite difficult to encourage and to cultivate without building understanding and trust among the stakeholders. Trust is the core of the process of breaking down stereotypes and other barriers and it helps to build mutual respect and creates a commitment to the process.

5.2 *Vision, planning and strategy*

It is difficult to achieve constructive planning and building shared vision with conflict among partners. Conflict needs to find a solution that satisfies stakeholders and this leads to deal successfully with the conflict. In this collaborative and cooperative model, it proposes to handle the conflict through avoidance of or position-based competition. In this approach, stakeholders in conflict do not deal with their differences and they hold to their positions and try to prevail their power over the other fragile one. The problem of this competitive approach is that one stakeholder win and the other lose. As a result, relationships be weakened and commitments to consensus in decision making may be weak. Thus, this model proposes interest-based problem solving approach. This approach means that there is a margin for stakeholders who have conflict to sit down together and to find a solution by dialogue about their mutual concerns and interest. The stakeholders feel more likely that the decision-making process has been a fair one and they tend to be more committed to carrying out the agreements made.

Commitment relays on trust and respect for stakeholders' different perspectives and interests. Commitments form clear, transparent, and fair procedures in the right decision-making process. Additionally, commitment creates mutual confidence among stakeholders that the procedures have integrity and fairness.

The researchers suggest that good planning and cooperative and innovative vision and strategy are more likely to ensure when the possible purposes and advantages of collaboration are relatively concrete and when win-win approach gained from collaborative-collective decision making.

5.3 *Win-win policy making*

Successful and productive outcomes of the collaborative process may represent tangible outputs and practical evidence for the stakeholders and it keeps up the progress and the momentum that leads to sustainable and durable cooperation. Win-win outcomes allows all the stakeholders to gain from their decision-making and encouraging them towards continuing working together towards finding solutions to their barriers that result in brining satisfaction for partners. The win-win atmosphere maintains the relationship among different partners and encouraging them to focus on interests not positions, rights, or stereotypes. This approach equips different partners and helps them to generate a variety of options and solutions based on an objective and interest-based standards which offer gains to the stakeholders.

6 CONCLUSION

Shifting collective decision making sphere to collective-collaborative governance model refers to the process in which the stakeholders work as a team in a cooperative-collaborative way in terms of governance system. The process expands the democratic participation, increases the benefits, and reduces the management and decision-making costs. The continuous process of collaborative and cooperative governance gives the opportunity for the stakeholders to engage in productive dialogue, develop a positive and fruitful relationships, and equip the participants with a variety of long-life learning skills such as; problem solving, team work, negotiation, bargaining, planning, fundraising, and commitment to public interests.

Based on an experience analysis of Halland model and a strategic analysis of different historic conservation projects within Halland project, our findings are largely empirically inductive. In analysing these projects, our goal has been to identify the opportunities that consolidate and facilitate the process of the trading zone collaborative decision-making venue.

Professionals in historic conservation and planning agencies can use the development process of The Halland Model to identify key challenges and limitations to their collaborative planning and decision-making process.

Halland model has viewed as a catalyst for sustainable development, and the model states that utilizing "trading zone" as a democratic pluralistic arena in historic conservation's decision-making process is the catalyst for sustainable regional development. The model developed where the "trading zone" defined as an active melting pot aiming at assisting various actors and stakeholders including; developers, conservators, public sectors, and public representatives for making sense together.

However, in this paper we conclude by emphasizing that the governance process of implementing trading zone philosophy was missed, especially, those variables which related to the durability and governance factors of public sphere. This paper suggests practical process aims at making enhancement for the model in order to achieve an integrated vision of historic preservation within regional development planning process. In this way, collective (cooperative-collaborative) governance sphere "the enhanced model" is constitute by the stakeholders and the members of an intercommunicating community, situated in the particularities of time, place (value and significance), and regulations (historic preservation, development, and planning) aiming at guiding and managing sustainable long-term actions. The enhancement of Halland model shows not only how to make sense and taking decisions together, but also working out how to act together.

REFERENCES

Amiry, S. (2004), Job Creat paper presented at *Swedish Urban Heritage Network at World Urban Forum II*, Barcelona, Spain 2004.

Balducci, A. and Mäntysalo, R. (eds) (2103). *Urban Planning as a Trading Zone*. Springer press. ISBN-13: 978-9400758537.

Ferilli, G., Gustafsson, C., and Sacco, P.L. (2017). *Cognitive Keynesianism: Heritage conservation as a platform for structural anti-cyclic policy. The case of Halland Region, Sweden*. In: Journal of Cultural Heritage.

Galison, P (1997). *Image and Logic: A Material Culture of Microphysics*. University of Chicago Press, Chicago.

Gustafsson, C., (2011). *The Halland Model: A Trading Zone for Building Conservation in Concert with Labour Market Policy and the Construction Industry, Aiming at Regional Sustainable Development*. Second Edition. Gothenburg: Gothenburg Studies in Conservation, Vol. 24.

Lasker, D., Weiss E.S., and R. Miller R. (2001). *Partnership synergy: A practical framework for studying and strengthening the collaborative advantage*. The Milbank Quarterly, Vol. 79, No. 2, pp. 179–205.

Muñoz Viñas, S. (2005). *Contemporary Theory of Conservation.* Oxford: Elsevier Butterworth Heinemann.

Sörlin, S. (2001). "The trading zone between articulation and preservation: Production of meaning in landscape history and the problems of heritage-decisions-making. In: Baer, N.S. and Snickars, F. (eds.) *Rational Decision-Making in the Preservation of Cultural Property.* Berlin: Dahlme University Press.

Yin, R.K. (2014). Case Study Research. Sage Publications.

Innovative Built Heritage Models – Van Balen & Vandesande (Eds)
© 2018 Taylor & Francis Group, London, ISBN 978-1-138-49861-7

Cultural heritage as a resource for regional sustainable development: The example of the Valtellina Cultural District in Italy

D. Foppoli
Foppoli Moretta e Associati, Tirano, Italy

ABSTRACT: The Cultural Districts project was developed over the 2005–2010 period by the Cariplo Foundation to support the enhancement of cultural heritage as a flywheel for economic growth and social promotion of territories through the promotion of the integrated development process of cultural heritage. Following the guidelines issued by Cariplo Foundation, a number of specific communities in Lombardy worked to build a Cultural District in their territory. The project of Valtellina Cultural District was deemed worthy of being funded and so the District was established in 2010. The aims of the project involve cultural landscape, heritage, awareness of local community, strategical development forecast and creation of tangible or intangible networks. To pursue these goals, the district has developed specific work packages aimed at restoring heritage buildings, Preventive Planned Conservation, skills/capacity enhancement, community involvement and valorisation. The case history is discussed through the analysis of the partnership and of the methods with which the partners have agreed upon the idea that cultural heritage is a resource for sustainable development of their territory. For this purpose examples of relevant activities are presented, such as "The Route of Terraces", that is one of the District's most suitable work packages for creating networks and stimulating the link between local population and cultural landscape. The awareness of the extent of the resources of the territory generated by the Cultural District led to the evolution of the project in an EU cross-border cooperation process that takes advantage of the historical connections (both cultural and administrative) between Valtellina and Canton Graubünden in Switzerland. This also allowed for the extension of the partnership through the involvement of new partners coming from the nearby Swiss territory. The analysis is aimed at underlining strengths and opportunities provided to the territory and weaknesses and threats inherent to the development process, but also to point out the outcomes of the project.

1 INTRODUCTION

Cultural heritage is a genuine driver for the social and economic growth of the territory but only where it establishes an actual network connecting tangible and intangible assets and implementing political, socio-economic and legal frames. This consideration is the starting point for the development of the Cultural Districts experiences that in Italy can be traced back to three main typologies: regional initiative, provincial initiative, and banking foundations initiative (Cerquetti & Ferrara, 2014).

Cultural Districts project proposed by Cariplo Foundation is a wide area project fully included within the third typology. It is applied to the Italian territory of Lombardy and is split into six subprojects; one of them, Valtellina Cultural District, is the specific argument of this paper. This case study has been analyzed, particularly highlighting in which way it has pursued and obtained in its execution the maximum involvement of all the stakeholders; a SWOT analysis was also carried out and some relevant outcomes of the project are outlined, mostly in terms of capacity building.

1.1 *Cultural Districts project*

Fondazione Cariplo (Cariplo Foundation) is an Italian private grant making foundation with a banking origin. It was established in 1991 and it is one of the world's main philanthropic organizations. It aims at operating on the basis of the principle of subsidiarity, foreseeing needs and acting to help social and civil organizations to better serve their community. For this purpose it implements its own projects bringing partners together and catalyzing resources.

The Cultural Districts project was developed to promote economic growth in Lombardy valorising the local cultural heritage. The term 'Cultural District' denotes an area rich in cultural and environmental assets, industrial and service businesses that operate in a synergical manner (Barbetta et al., 2013).

The project is set out in four steps.

1st step (2005–2006): a general pre-feasibility study was conducted to enable Cariplo Foundation to identify areas with similar features that could potentially become Cultural Districts. It was carried out by Milan Politecnico university on the entire territory of Lombardy (with its neighbouring areas) and was aimed at setting up the basic operational criteria. This led to the identification of 34 homogeneous areas in which the districts might find suitable conditions to be developed.

2nd step (2007–2008): a call for proposals was launched and, as a result, 11 potential cultural districts were identified. Their institutional framework and economic sustainability were analyzed in-depth to determine their viability as cultural districts.

3rd step (2008–2010): local teams in the 11 selected areas worked on the development of operative feasibility studies to create a cultural district with the support of the foundation that provided them with guidelines and the supervision of a panel of experts. Cariplo Foundation reviewed the outcomes of these studies and selected 6 areas suitable for the development of a Cultural District. It funded each of these areas with up to € 3.8 million and matching funds were provided by other local sources.

4th step (2009–2010) start-up of the activities of the Cultural Districts in Lombardy which were established in the following areas: Camonica Valley, Mantua Po River Area, Gonzaga Palaces Area, Monza & Brianza Province, Cremona Province and Valtellina.

In the following article, the case history of the Valtellina Cultural District will be presented and discussed.

2 VALTELLINA CULTURAL DISTRICT

2.1 *Previous experiences in Valtellina*

The awareness of the feasibility of a Cultural District in Valtellina was based on the experience of managing complex processes concerning cultural heritage that was gained by the territory over the last twenty years.

The first organic approach to conservation activities of cultural heritage in this area dates back to a catastrophic event: the 1987 flood. Following this event, the reconstruction was funded under a special law (L. 102/90 then called "Legge Valtellina") which also provided resources for the restoration works of more than 70 heritage buildings. A total amount of € 35 million, financed with 25% own resources, was necessary to carry out these works. Although the results were positive, the outcomes of the operations were not subjected to a good approach with methods of Preventive Planned Conservation (also PPC): consequently, in spite of the relevant investments, the accomplished work has not provided effective economical and social growth for the territory.

Over the years following 2002, in some cases even with the support of Cariplo Foundation, many long-term projects have been developed with an improved approach to PPC methodologies.

The first reference is to the call "Promoting of Cultural Heritage through the Integrated Management of Assets", that funded the project "Antica Pieve di Mazzo" (2006–2009) for a total amount of € 1,991,000; this was funded for 41% by own resources (Foppoli D. et al. 2008). Then the call "Innovative Technologies for Preventive Conservation of heritage Buildings" that funded the project "Palazzi Besta" (2009–2010) for a total amount of € 345,000, funded for 44% by own resources (Armanasco & Foppoli, 2014).

It is of interest to underline the fact that the co-financing by own resources has increased over the years from 25% for the project funded in 1997 to 41–44% for the project funded in 2006–2010 to 56–57% for the last projects that were funded in 2011–2015, that will be presented in the next chapters.

This is the result of the current economic crisis affecting the Italian public administration, but it is also a signal of the increasing involvement of the territories in the projects.

2.2 *Cultural District project*

The activity of Distretto Culturale della Valtellina (also DCV—Valtellina Cultural District) started in 2009 when the local team was promoted by the provincial administration and the operative feasibility study was developed. The local team was the result of a good integration of experts of several disciplines including economic, statistic, historical, architectural, landscape, legal and administrative aspects

A number of specific questions were placed at the beginning of the study, starting from the most suitable extension of the district: initially it was proposed a further division of the territory at the boundary of the four Mountain Councils (Morbegno, Sondrio, Tirano and Alta Valtellina) which already exist as administrative units. Further studies underlined that only by considering the whole Adda Valley was it possible to best exploit the enormous potential for tourism in the Alta Valtellina. On the other hand, the cohesion carried to the territory from the terraces that characterise the landscape of the valley proved to be a catalyst for the creation of the partnership of the Cultural District.

The overall surface of the district is 2.760 km^2, the population is 155.000 (it is necessary to underline that this is a mountain area with a non-homogeneous distribution of the population over the territory): it includes 4 mountain councils and 65 municipalities.

The operative feasibility study was developed paying attention both to content and to economic resources, with a particular regard to activate specific tangible interventions with the characteristics of economic feasibility, representativeness and capacity to create tangible and intangible networks over the territory.

The "Valtellina Cultural District" Project (2011–2015) had a total funding of € 8,290,000 and was funded partly by Cariplo Foundation and for 57% by own resources coming from the territory.

2.3 *Partnership*

The partnership of the DCV was promoted by Sondrio provincial administration, including municipalities (Morbegno, Sondrio, Tirano, Valdidentro), mountain councils (Morbegno, Sondrio, Tirano, Alta Valtellina) and others boards, associations and foundations involved in the development of local economy, tourism and culture. The project also involves funders such as Sondrio Chamber of Commerce, local banks, an electric power production company and some local administrations of the nearby Swiss territories. The project leadership was assigned to Fondazione di Sviluppo Locale (also FSL—Local Development Foundation), which is partnered by all the partners and funders of the project.

The Valtellina Cultural District was finally established in 2010. In agreement with the purposes of Cariplo Foundation, it aims to support the opportunity to enhance cultural heritage as a flywheel for economic growth and social promotion of Valtellina and to promote integrated development process of cultural heritage, of an economic nature too.

This overall goal is specifically laid out in the following objectives:

- promoting both inward and outward knowledge of Valtellina cultural heritage;
- integrating products and cultural services making them attractive, economically and environmentally sustainable and oriented to social-economical development;
- increasing outwards visibility and reputation of Valtellina and of its products;
- improving cultural management, mobilizing cultural growth potential and contributing to employment and welfare creation;
- developing a shared and integrated modality of programming and planning to manage, in medium and long term, further actions which are organic and coherent with the themes of the Cultural District.

The project develops specific work packages (also WP) aimed at restoration works, Preventive Planned Conservation, maintenance activities, skills/capacity enhancement, community involvement and valorisation.

2.4 *Strategies for partner involvement*

A crucial factor is to evaluate how the partners have agreed about the fact that cultural heritage is a resource for sustainable development.

To this purpose, it was necessary to create networks and develop complex activities that however were transversal to the interests and objectives of each of the partners. In this respect, tangible activities aimed at preserving cultural heritage and landscape have been a powerful tool as they seek analysis and reflections and generate externalities that are able to stimulate multiple and converging interests.

Figure 1 shows the operating scheme of one of these WPs, which highlights the objectives of the project and the tools that allowed the involvement of such a great number of partners.

In this case the experience demonstrates that through significant collective efforts it is possible to draw attention to the territory and to sustain its recognition as a cultural landscape. Therefore, the awareness of the historical and economic significance of the landscape supports the acknowledgement of the key role played by conservation and maintenance in actively protecting the territory from environmental and geological damage.

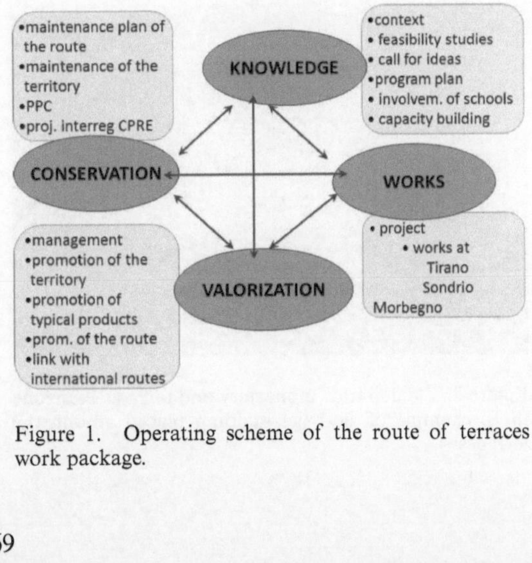

Figure 1. Operating scheme of the route of terraces work package.

3 EXAMPLES OF WORK PACKAGE DEVELOPED

3.1 *The route of terraces*

The main WP carried out by the DCV involved the creation of a route aimed at the valorisation of the terraced landscape on the slopes of the Rhaetian Alps in Valtellina. Here the vineyards were obtained by terracing the steep, rocky mountain slopes using dry stone walls (Figure 2): this terraced district is the largest wine-growing mountain terraced area in Italy, extending over 850 ha and representing 37% of the total Italian terraced vineyard surface.

La Via dei Terrazzamenti (The Route of Terraces) which has been designed foresaw the linking up and use of several already existing routes placed along the mountain slopes, thus creating a trail that boasts cultural sites, with constant reference to local agro-food production activities. Thereby, over the long term, fostering the recovery and valorisation of major and minor cultural heritage sites, creating an authentic tourist attraction within the overall cultural, tourist, excursion and recreational network offer (Di Capita & Foppoli 2016).

The initiative was entrusted to FSL and supported, funded and implemented by a variety of bodies representing many of the public and private stakeholders operating in Valtellina. In order to create effective synergy, all the subjects identified as stakeholders were involved in designing and planning the route.

The way of developing the activities provides an example of the method and, especially during the programming and planning phase, fostered the maximum degree of involvement of professionals with specific expertise who are present in the territory. The explicit aim was to overcome conflicts of interest and to adopt a collaborative approach, which is the best way to really pursue synergy.

Figure 2. "La Gatta" monastery and farm at Bianzone (SO), example of heritage building placed among the terraces.

Furthermore, particular attention was paid to educational activities, with multi-purpose objectives (landscape—cultural heritage—Planned Preventive Conservation), targets (professionals—teachers—students) and approaches (courses—workshops—competitions).

Another aspect to be emphasized is the great importance of management: each one of the stakeholders agreed that The Route of Terraces should be "managed" not only by the involved territorial authorities, but also and above all by that part of the local community that profits from the territorial development. These subjects, who were involved through their specific trade associations, fully approved the initiative and showed keen interest and appreciation for the project. They readily disseminated information on the project among members of the associations, fully recognizing that a joint strategy could be not only extremely effective, but also advantageous for all the parties involved. The action that was implemented was aimed at leveraging the sense of belonging and identification with the culture and landscape of the area. The goal is to promote the quality, the management and maintenance of the Route through effective choices in terms of contributions (not only of an economic nature, but also with practical maintenance activities) by the economic operators.

Due to the position of the Valtellina—on the border between Italy and Switzerland—particular attention was paid, within the framework of promotional activities, to create a link with the experiences developed in nearby Switzerland. From the start of the project, "The Route of Terraces" project was submitted to "Via Storia", the association that promotes Swiss historical routes, which decided to include it as an extension of the existing Via Valtellina (the route whereby Valtellina wine was historically exported to the Northern territories of Europe), which pass though the Canton Graubünden (in Switzerland), reaching Voralberg Region (in Austria).

Therefore, it was possible to track a route that is closely linked to the culture of production and commerce of the Valtellina wine, linking Italy, Switzerland and Austria, which in the future could connect Bodensee with Como Lake, becoming a relevant tourist and cultural attraction for the involved territories.

3.2 *Conservation projects*

Conservation projects developed under the DCV concerned both the restoration of some buildings of particular relevance and the large scale assessment of the state of conservation of cultural heritage buildings restored with the funding of L 102/90 as mentioned above.

For most of these buildings, at the time of the restoration, a proper inspection and maintenance plan has not been implemented; the intervention carried out by DCV was aimed at both assessing the effectiveness of the restoration techniques applied during the works and evaluating the effects of non-maintenance over time.

For each of the involved buildings, a detailed inspection report was drawn up and also the assessment of the seismic vulnerability was carried out at the preliminary level (LV1). This last activity, indeed, had not been carried out at the time of the interventions because it had not yet been introduced in Italian structural code.

With regard to seismic assessment, it has to be noted that the buildings analyzed presented a generally reassuring status, especially with reference to the low seismicity of the area, but highlighted some criticalities too, especially related to the geometric discontinuities between the nave of the churches and the bell-towers.

The correct approach to conservation issues has also been carefully pursued by DCV in the interventions on cultural heritage buildings, which have been made within the specific WPs. They concerned S. Antonio Cloister at Morbegno, Masegra Castle at Sondrio and Tirano Railway Station; the last building was selected because it is included within the buffer zone of the transnational UNESCO site "Rhaetian Railway in the Landscape of Albula—Bernina".

The proper approach to the interventions required for these buildings included:

- a preliminary selection based on the possibility of creating tangible or intangible framework on the territory;
- the predisposition of management plans from the beginning of the feasibility studies;
- the carrying out of accurate preliminary diagnostic analysis to gain the significant cognitive apparatus necessary to have full knowledge of the building;
- the execution of the carefully planned works, monitored by the DCV actuators (municipalities and FSL);
- the definition of conservation plans (Bossi et al. 2016) to support the future maintenance and valorisation processes.

Since the interventions carried out by the DCV concerned very large buildings, in every case they could not involve whole complexes; they were anyway very significant because they helped to draw up a road map (and above all an operating model). In the cases of Castel Masegra and Tirano Railway Station, this has subsequently led to further stages of intervention and valorisation.

3.3 *EU Cross-border cooperation project*

The aforementioned high-profile analysis and research projects developed within the Valtellina Cultural District in the field of conservation and valorisation of cultural assets have enabled positive exchanges with neighbouring Swiss territory. They have led to the elaboration of the project "Preventive Planned Conservation in the Rhaetian Common Space—CPRE Project", which was an immediate outcome and development of DCV project. It became operational through the funding of the EU 2007–2013 cross-border Italy-Switzerland inter-regional cooperation program (Foppoli et al. 2014)

The aim of this project was to enhance the close links between Valtellina and the Graubünden Canton (in Switzerland) in order to coordinate the conservation of the architectural and landscape heritage of Valtellina and Valposchiavo, sharing projects, knowledge and methodologies. The total cost of the project (developed over the years 2013–2015) is € 1,570,000, funded for 56% by own resources.

The goals of the project can be summarized in four macro-themes: knowledge, conservation, training, valorisation.

The knowledge involves history, materials, constructive techniques, but also, and above all, the context in which cultural heritage is placed and the correlation between elements of different meaning/function. The conservation has been developed on the basis of a systemic approach and long-term vision, which seeks to interact positively on the UNESCO heritage site, receiving stimulus and providing operational proposals.

Special attention has then been paid to the integration with DCV projects on the theme of capacity building, developed with the involvement both of technicians (architects, engineers, restorers) and operators (craftsmen and workers). These projects have common goals and an integrated and multidisciplinary approach with the aim of providing both sides of the border with shared competences that can be invested in their respective territories.

The training of technicians involved courses for the definition of specific professional profiles, both on the theme of Preventive Planned Conservation and of the analysis of the cultural landscape. The training of the craftsmen was aimed at the activation of modular courses that provide a certificate of professional recognition with validity both in Italian and in Swiss territories. It should be noted in this regard that, in this area, the number of cross-border, annual or seasonal workers operating in the construction sector is high and with these courses they can thus obtain well-qualified certifications that can be useful both in Italy and in Switzerland.

Surely, the development of the inter-regional project in synergy with the DCV has enabled a successful exchange of experiences to create a centre of competence enhancing the two adjacent territories and allowing them to gain a cultural centrality in the Alpine geographical region.

4 RESULTS AND OUTCOMES

4.1 *Helpful and harmful elements highlighted by the experience*

A SWOT analysis of the activities of DCV is carried out in Figure 3. The positive outcomes of the experience are based on the opportunities provided by the territory in which the District was located. In Valtellina there is a strong and clear integration between cultural heritage and landscape and the territory is coherent: the people recognize their identity, although sometimes in a not entirely defined manner. The territory has proved to be ready to process complex development projects mainly thanks to the experience gained from previous projects.

Local and regional authorities are available to work together, although mostly on tangible assets; furthermore, the project was developed with a strong proactive involvement of the public authorities competent for cultural heritage: Italian Superintendence of Architectural and Cultural Heritage and also Swiss Denkmalpflege. These are some of the most relevant key elements of the success of the developed projects.

	HELPFUL	HARMFUL
INTERNAL ORIGIN	**Strengths** •clear methodological guidelines provided by Cariplo Foundation •tangible goals are powerful tool for partners aggregation •partnership tried to be as inclusive as possible •project involved widely shared training activities	**Weakness** •territory has struggled to recognize the endogenous quality of this project •tangible assets has sometimes overshadowed intangible assets •difficulty involvement of partners in a long-term planning no resources for the start up
EXTERNAL ORIGN	**Opportunities** •experience gained from other previous projects •strong and clear integration between heritage and landscape •people recognizes their identity •authorities and supervisory boards are available to cooperate	**Threats** •the actual touristic model is nowadays outdated because based on skiing and moliday homes •lack of effective political representativeness •administrative upheavals of recent years

Figure 3. SWOT analysis of Valtellina Cultural District project.

On the other hand, in Valtellina there is a touristic model based on skiing and on holiday homes that nowadays is obviously outdated. This consideration is shared, but large scale capital is yet to be invested in this direction, rather than in other innovative and sustainable sectors and this constitutes a significant obstacle to the projects developed in culture activities such as heritage valorisation.

This was the reason the territory proved not to be entirely homogeneous: the powerful tourist attraction provided by Alta Valtellina produces social and economic dynamics on which it is difficult to operate. Indeed, it must be acknowledged, that the involvement of the Bormio Mountain Council area in the project was inappropriate. On the other hand major but not yet fully consolidated touristic potential, such as those provided by UNESCO in the territory of the Tirano Mountain Council, could be well integrated into the DCV project. Furthermore, this was the lever, from which the involvement of Swiss territories in the district's projects began.

It was helpful for the development of the project that it was based on very clear methodological guidelines provided by the Cariplo Foundation, suitable to pursuing a wide and adaptable grid of goals that were defined by local teams. Experience has shown that the territory initially struggled to recognize the endogenous quality of this project, especially because of opposition in the local cultural sector. This problem was overcome with wide ranging presentation and sharing of the project and with the right operations to enlarge the partnership trying to be as inclusive as possible. It was also important to increase the reputation of the project through building capacity activities that have been well shared among stakeholders.

Tangible goals proved to be a powerful means of aggregation of local interests around the project, but the importance of investment on tangible assets has sometimes overshadowed the need to operate in a robust manner on intangible assets as well.

A relevant weakness of the project was the local and general political situation. Firstly, there was lack of effective representation from the people institutionally representing the territory: this makes it difficult to recognise the true needs and the available resources to properly implement the project. Also, it was difficult to involve local political representatives in broad and long-term planning. Moreover, substantial difficulty was experienced due to the administrative upheaval that has occurred in Italy over recent years, such as the proposed abolition of the provincial administrations that has caused the gradual disappearance of the territorial authorities of reference that initially sustained the start up of the District Project. Finally, it was possible to foresee and overcome this problem by the settlement of the Local Development Foundation, that is an independent entity appointed as project leader.

Another weakness in project planning, pointed out by comparison with the projects of the other districts, was the lack of capacity to involve young people in creativity and entrepreneurship. This aspect, which had not really been foreseen, was not implemented in the DCV thus proving limiting in terms of vision of future assets.

A relevant difficulty for the project was finally due to the fact that the Cariplo Foundation has invested huge resources, but it has not provided additional resources to ensure the effective start up of the valorisation projects after the initial three years.

4.2 *Outcomes of Cultural Districts project*

Regarding merely economic terms, the Cultural District projects have been a powerful lever for the economic growth of Lombardy: over 5 years, the Cariplo Foundation granted € 19 million to produce a global investment, for the 6 districts, of more than € 58 million with the involvement (in the production and management of culture related projects) of more than 3,000 suppliers. This was a useful and powerful element in this period of general economic depression in Italy.

DCV was a part of this effort and allowed an overall investment on the territory of Valtellina of more than € 8 million that led to the creation of a 70 km touristic route, to the restoration of 3 heritage buildings, to the carrying out of courses and conferences on heritage and landscape with more than 3,000 participants.

Dealing with the initial purposes of the project it can be stated that DCV was very effective in promoting knowledge of cultural heritage, in integrating cultural landscape and local products (especially agro-food ones) and in increasing visibility and reputation of Valtellina, as proved by the good results of several specific WPs, for example the previously described "The Route of Terraces".

On the other hand the activities of the District were not so effective in pursuing the economical sustainability for all the cultural services and in generating employment and welfare opportunity: for example the planned valorisation of S. Antonio Cloister is not currently running, due to the lack of long-term vision and of economic resources after the start-up of the activity.

With regard to the development of further steps actions coherent with the themes of the District, the most important result has been the creation of figures able to activate further projects on the territory: over the last years, after the completion of the initial phase, a number of further projects linked to key-words of DCV operational activity were proposed.

Figure 4. Maintenance activities carried on by recently unemployed workers under the project "+++ More Positive Signs".

In 2014, under the Cariplo Foundation call "Welfare in Action", the project "+++ More Positive Signs" was set up. It was a three-year project planned by a network of both public entities and social cooperatives and focused on the "maintenance" of paths, terraces and landscape. The project came from the idea that it is possible to bring together two important weaknesses of mountain territories (both social and hydro-geological weaknesses—Figure 4) to produce a relevant strength (opportunity of work in maintenance).

Another key-word that has been retrieved over recent years is "valorisation": Tirano Mountain Council, in partnership with all the municipalities of the territory, has planned the project "Conservation and Valorisation of the Cultural Landscape of Mid-Valtellina". This is a project of broad scope, presented to the Cariplo Foundation for a grant, that will fully implement the Route of the Terraces through the proposal of high profile conservation and valorisation projects for the terraced areas.

The last relevant is "Cultural Heritage and Cultural Landscape of Valtellina and Valposchiavo in the UNESCO Context", that Tirano municipality is planning to continue and develop the themes of CPRE project under the grant of the current EU Cross-border cooperation (2014–2020).

5 CONCLUSIONS

Following the idea promoted by the Cariplo Foundation a Cultural District was settled in Valtellina to sustain cultural development and economic growth of the territory through projects of valorisation and the creation of networks.

Valtellina proved to be a community with a strong identity, a territory with a good integration between cultural landscape and heritage buildings and with significant previous experience of networking activities in this field.

This is why it was possible to develop the Valtellina Cultural District project through the creation of a coherent partnership. This set out the definition of clear, sustainable and shared goals and the development of specific work packages aimed at exploiting tangible and intangible assets.

The project reached its objectives and produced interesting outcomes; there were, however, some difficulties due to administrative upheaval and especially to the long start up time that is required for such a complex project, that proved to be longer than that initially foreseen by the Cariplo Foundation.

REFERENCES

Armanasco A. & Foppoli D. (2014). Diagnostic tests vs structural models: the utility of the comparison, Proceedings of 11th International conference on non-destructive investigations and microanalysis for the diagnostics and conservation of cultural and environmental heritage (Art'14), Madrid.

Barbetta G.P., Cammelli M., Della Torre S. (a cura di) (2013). Distretti Culturali: dalla teoria alla pratica, Bologna: Il Mulino.

Bossi S., Caligari A., Moioli R. (2016). A sustainable management process through the preventive and planned conservation methodology: the Conservation Plan of the complex of St. Anthony, *Proceedings of the 5th International Conference on Heritage and Sustainable Development (HERITAGE 2016)*, Green Lines Institute, Rogério Amoêda, Sérgio Lira & Cristina Pinheiro.

Cerquetti M., Ferrara C. (2014). Distretti Culturali: percorsi evolutivi e azioni di policy a confronto in Proceedings of workshop "Distretti culturali: esperienze a confronto", *Journal of the section of cultural heritage*, University of Macerata.

Della Torre S. (2014). Lezioni imparate sul campo dei distretti culturali in Proceedings of workshop "Distretti culturali: esperienze a confronto", *Journal of the section of cultural heritage*, University of Macerata.

Della Torre, S. (2015). Shaping tools for built heritage conservation: from architectural design to program and management. Learning from "Distretti culturali". *In: Van Balen, K. & Vandesande, A., eds. Community involvement in heritage management*. Antwerp: Garant.

Di Capita F. & Foppoli D. (2016). The Route of Terraces in Valtellina: Community involvement and tourism for the enhancement of cultural landscape, *Proceedings of 3rd World meeting on terraced landscape, Venice - Padua*, in printing.

Foppoli D., Menghini G., Zanolari E. (2014). Preventive conservation: an opportunity of cooperation in the heart of the Alps, *La strategia della conservazione programmata. Dalla progettazione delle attività alla valutazione degli impatti - Vol. 1*, Milano: Nardini.

Foppoli D., Giudici P., Zucchella A. (2008). I Beni Culturali come strumento di crescita per il territorio: Il progetto Pieve di Mazzo, *Proceedings of conference: Cultura e Territorio - Beni e Attività culturali: valorizzazione e indotto in prospettiva europea*, Pavia: Franco Angeli.

Innovative Built Heritage Models – Van Balen & Vandesande (Eds)
© 2018 Taylor & Francis Group, London, ISBN 978-1-138-49861-7

Sustainable strategies in built heritage preservation: Planned conservation and community participation. Experiences in Portugal

T.C. Ferreira
Centro de Estudos de Arquitectura e Urbanismo, Porto, Portugal

ABSTRACT: In the present-day context of a sharp decrease in economic and ecological resources, preventive conservation is a key strategy for sustainable heritage management, because of its cost effectiveness, increased preservation of authenticity and community driven development. Nevertheless, there are still very few real implementations, so that the purpose of this paper is to illustrate Portuguese experiences with special focus on the Rota do Românico, a Cultural Route with 58 monuments (monasteries, churches, towers, bridges, chapels, memorials) over an area of 764 km^2 and involving 12 municipalities. This case study demonstrates that planned conservation is increasingly successful when linked with the participation of local communities and users.

1 INTRODUCTION

1.1 *Context and concepts*

The current context of economic depression, ecological saturation and an excessive number of recent constructions (many of which are empty, useless or obsolete) calls for a careful approach to the management and preservation of the built heritage. Acknowledging that the heavy and exceptional restoration of single monuments is no longer an exclusive or sustainable practice, it is urgent to implement strategies for preventive conservation and maintenance, which have always existed in the history of construction (plastering, replacing tiles, cleaning gutters, etc.), especially before its industrialisation. These practices provide for a more sustainable management of resources, shifting from a reactive and interventionist approach (post-damage) to preventive conservation (pre-damage) and continued care over time.

Seen from this perspective, attention can be drawn to the work of some international organizations and to initiatives fostering planned and preventive conservation, including, among others, the Society for the Protection of Ancient Buildings – SPAB, in the UK (Crofts, 2008), the Monumentenwatch in Belgium and The Netherlands (Vandesande & Van Balen, 2016), the Planned Conservation experiences in Lombardy, Italy (Della Torre et al, 2003), the Monitoring System for the World Heritage Site of Cuenca (Heras et al, 2013) as well as the commitment of the UNESCO Chair in Preventive Conservation, Maintenance and Monitoring of Monuments and Sites – (PRECOM^{3}OS) (Van Balen & Stulens, 2001; Van Balen & Vandesande, 2013). There is a vast lexicon used in the field of preventive conservation and maintenance, which also varies according to different languages, countries and fields of action (Forster & Kyan, 2009; Lipovec & Van Balen, 2010; Ferreira, 2011a, Vandesande, 2017).

In Portugal, the term "maintenance" is more commonly used, broadly meaning the "set of preventive operations carried out in order to keep a building as a whole or each of its constituent parts in a good state" (Henriques, 1991). Maintenance is thus a system of preventive practices, controls, small repairs, cleaning and appropriate uses. In this sense, one can identify different types of maintenance: preventive or predictive maintenance (inspection and preventive actions), conditioned maintenance (post-symptomatic actions) and curative or corrective maintenance (post-damage actions) (Paiva et al, 2006; Calejo, 2009; Flores-Collen & Brito, 2003). Maintenance is therefore a broad term, encompassing a vast range of different operations, which may be direct or indirect, simple or qualified, varying also in their frequency (daily, weekly, monthly, biannual or multiannual) or according to the seasons of the year (Coias, 2004).

In this framework, preventive-planned conservation and maintenance are key strategies for ensuring sustainable built heritage preservation, with great potential existing for their application not only in single buildings, but also in public spaces, the landscape and the whole territory. However, there are still very few practical applications to real case studies, especially those linking preventive-planned conservation with the engagement and empowerment of local communities and users.

1.2 *Community participation*

Preventive-planned conservation and maintenance strategies when applied to the built heritage require a knowledge of traditional materials and techniques, which further enhance compatible and sustainable repairs. These works require a skilled and qualified labour force, which, in many cases, is quite aged or has been undervalued due to the effects of the industrialisation of construction. However, maintenance of the built heritage can be a way to foster local participation, professional qualification and employment, by reactivating materials and technologies that have fallen into disuse or are at risk of disappearing (ICOMOS, 2014, 5).

On the other hand, community empowerment calls for the participation of all users in preventive conservation and maintenance processes, namely with the support of "User Manuals" and "Instruction Manuals", which provide information and guidelines in an accessible language. Hence, users are a key factor in heritage maintenance, helping to avoid improper use, preventing risky situations, contributing to the recording of information and collaborating in daily maintenance activities (cleaning, ventilation, shade, etc.). From this point of view, civil society has an important role to play in preserving heritage, as well as in contributing to a sustainable and culturally integrated development (Ferreira, 2011): "Local people, civil society, and elected local and national officials will play a key role in the design and implementation of heritage as a driver of development, and, through raised awareness of heritage, they will have ownership of the development process" (ICOMOS, 2011, p. 6).

1.3 *Objectives and methodology*

The aim of this paper is to present research applied to real case studies in Portugal, as well as to demonstrate that planned conservation is increasingly successful when linked with the participation of local communities and users. Moreover, this paper intends to present economical data supporting the advantages of preventive-planned conservation strategies, as well as to provide for operative tools, such as a computer software/App and illustrated manuals for users.

Moreover this paper proposes a methodological approach—(1) Preventive-planned conservation (material component)—performed directly on buildings, through inspection, monitoring, maintenance and repair activities; (2) Community empowerment (intangible component)—undertaken indirectly on users, through participatory and empowerment strategies. Besides, the article presents different case studies, from different epochs and with different functions, with the objective of supporting evidence and the results. In order to provide more detailed information, while increasingly supporting the methodological approach defended in this paper, special attention will be given to the case of the Rota do Românico. The following experiences are among the first experiences of preventive planned conservation strategies applied to the built heritage in Portugal.

2 PORTUGUESE EXPERIENCES

2.1 *Window maintenance in Porto*

The workshop "JANELAS com TINTA têm muito + PINTA" ("Painted windows are much more fun", organised by ICOMOS-Portugal in partnership with CEAU-FAUP and IC-FEUP) was a successful participatory experiment in the maintenance of windows in Porto's Historic Centre (UNESCO Word Heritage since 1996), performed between the 10–12 September 2012 in the area of Passeio das Virtudes. The aim of the workshop was to focus on an essential constructive element for the authenticity and identity of this site—the window—which has been a recurrent target for replacement, due to a lack of maintenance habits or as a result of intrusive renovation projects.

The workshop was targeted to local communities and took place over several days, involving practical demonstrations and direct "hands on" experiences throughout the different stages, techniques, instruments and materials applied in the maintenance of windows. The whole process was compiled in an illustrated Good Practices Manual (Figure 1) and was also recorded in a video published on YouTube (Romão et al, 2012). Moreover, this experience enhanced the potential

Figure 1. Illustrated instruction manual for windows maintenance in the Historical Centre of Porto (Source: The author).

of engaging and empowering local inhabitants through their active participation in the conservation process, increasing their quality of life and self-esteem through their direct relationship with the built heritage.

2.2 *Maintennace and participation strategies at the Faculty of Architecture in Porto*

The building complex of the Faculty of Architecture of the University of Porto (FAUP) was built in 1985–1992 under design and direction of Álvaro Siza, and is now listed as buildings with a special interest in the Heritage Charter of the Porto Municipal Plan. The FAUP buildings were designed to house 500 students and are currently used by about 1100 students. The building has thus been forced to adapt to a more intensive use (it is in permanent use—24 hours per day and seven days per week). The everyday users of the FAUP spaces are employees, teachers, researchers, students and visitors (tourists or others).

Recently (2016–2017), works were performed for the maintenance and repair of the external envelope of the buildings (roofs, façades and window frames) [1]. This repairs made it possible to solve some problems arising from the natural aging process caused by the wear and tear and the building's exposure to atmospheric agents, since it had not previously been subjected to any consistent repairs since its construction (Ferreira, 2016). The works made it possible to reflect on the implementation of planned conservation and maintenance strategies on the Faculty of Architecture (Ferreira 2017).

On the other hand, focusing on the importance of the participation of users, an infographic poster was developed (Figure 2) containing three levels of recommendations, ranging from (1) "suggestions" (organisation and cleaning of classrooms,

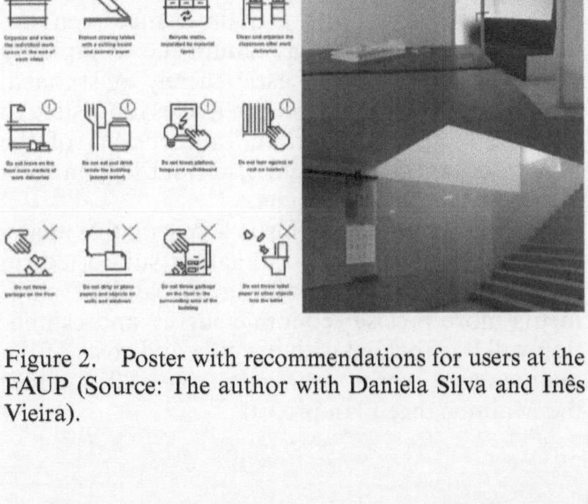

Figure 2. Poster with recommendations for users at the FAUP (Source: The author with Daniela Silva and Inês Vieira).

furniture protection, recycling), (2) to "alerts" (stowage of scale models, electrical wiring and equipment care) and (3) "prohibitions/obligations" (rules for the use of the sanitary facilities, the placement of litter and rubbish in their appropriate bins, writing or sticking posters on the walls and windows). This can be an incentive to increase awareness about the good use and preservation of spaces, until the time comes when such care is no longer a rule and becomes an automatic habit for all users (Silva 2016).

Only with the commitment and care of everyone – architects, managers, employees and users – can we ensure the proper functioning of the building, the preservation of its architectural heritage value and its sustainable transmission to future generations.

3 MAINTENANCE PLAN FOR THE MONUMENTS OF ROTA DO ROMÂNICO

3.1 *Context and objectives*

Rota do Românico is an exemplary experience as far as management and safeguarding strategies are concerned, benefiting from a network system and economies of scale (which are more attractive for visitors and investors) and seeking to strengthen local communities through an endogenous and integrated form of development (Ferreira 2011b). Hence, this project proposes a strategy that goes beyond the traditional (passive and regulatory) models of safeguarding to become a proactive system, generating economic synergies and the sociocultural development of the territory (Sacco, 2009).

The Rota do Românico covers 12 municipalities of the Vale do Sousa and Baixo Tâmega regions, comprising 58 monuments over an area of 764 km^2, including a range of different types, such as monasteries, churches, bridges, towers, memorials and a castle (Figure 3). Since 2003, it has proved possible to rehabilitate a considerable number of buildings, following good practices of intervention in the preservation of built heritage: (1) prior in-depth documentary and material surveys of the buildings; (2) diagnosis of the conservation status supported by non-destructive testing (NDT); (3) interventions using compatible materials and techniques, providing for the introduction of passive ventilation systems; (4) preventive conservation, monitoring and scheduled maintenance after rehabilitation interventions, operated through the Maintenance Plan.

The Maintenance Plan is an important feature of the Management Plan for the Conservation and Valorisation of the Monuments of the Rota do Românico [2], as it is a preventive conservation tool

Figure 3. Identification of Pilot Monuments, Rota do Românico (Source: Rota do Românico and the author).

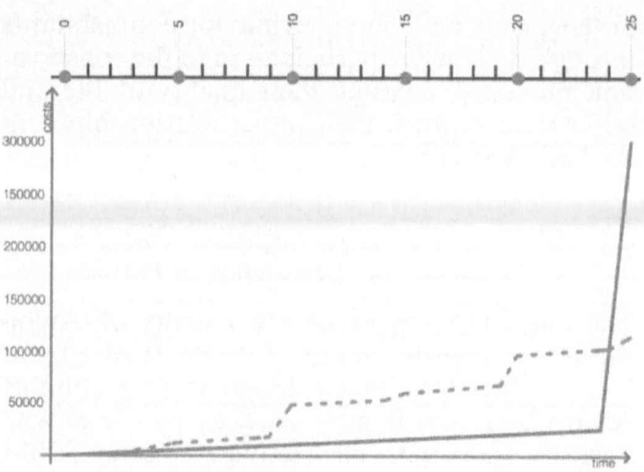

Figure 4. Chart comparing two scenarios of the accumulated costs of maintaining a religious building over 25 years: (1) without a Maintenance Plan (solid line); (2) with a Maintenance Plan (dashed line) (source: the author).

with the following objectives: to improve management of resources (reducing costs in the medium and long term), promote the future self-sustainability of the project through planned conservation and an appropriate management model based on preventive-planned conservation, avoid the risk or emergency situations, guarantee the quality of the tourism product by ensuring a good state of conservation of the buildings, promote employment and professional qualification, improve the self-esteem of the local populations through their empowerment and participation in the conservation process, and, finally, reinforce the pioneering and exemplary nature of the Rota do Românico in both a national and an international context (Ferreira 2012).

In order to better support the implementation of the Maintenance Plan, an hypothetic comparative analysis was undertaken for a church with a covered area of about 250 m² (Figure 4) which is the average area of a parish church of the Rota do Românico such as S. Vicente de Sousa. Thus, two scenarios were compared for the accumulated costs of construction work at that building, starting from a good condition point, over a 25-year time frame: (1) without the implementation of a Maintenance Plan (solid line), the costs are reduced over 25 years, after which a profound intervention will be necessary (€316,000); (2) with the implementation of a Maintenance Plan (dashed line), providing, in general terms, for regular interventions on an annual basis (visual inspection, clearing of vegetation in the surrounding environment, inspection and repair of roofs and other structural elements), every 5 years (painting of doors and windows as well as metallic elements, among others), every 10 years (cleaning of stonework, treatment of joints, preventive disinfestation of altars, structural inspection, among other interventions) with a total accumulated cost of €109,000.

This approach, although speculative, was important to persuade local stakeholders of the importance of funding planned conservation and maintenance strategies, also in terms of economic advantages. The detailed budgets were provided by an experienced contactor in conservation works (AOF).

This comparative analysis suggests that the total accumulated costs in the case of a building subjected to regular maintenance corresponds to about 1/3 of the costs of a profound intervention in the same building, which would be necessary after 25 years without these periodical maintenance works. In fact, in addition to the considerable reduction in costs, it is clear that it will be easier to obtain small sums of money for periodical annual and multiannual maintenance (in particular through the involvement of local agents, for instance, priests) than to secure large sums for profound and reactive interventions. On the other hand, it is also noted that, in percentage terms, annual costs (corresponding to only 0.46%) are relatively low when compared to the costs of a initial or profound intervention on the same building.

Moreover, even with a maintenance scenario, decay and deterioration cannot be completely avoided but, in some cases, merely postponed. However, regular maintenance makes it possible to increase the preservation and authenticity of the built heritage (heritage value), and subsequently its tourism and economic value.

Acknowledging that this is a speculative analysis based on hypothetic estimation supported on experienced contractor, in future research developments more precise economic survey and estimation will be provided with the data and costs of the maintenance works conducted since 2014 under the Maintenance Plan project.

3.2 *Methodology and structure*

The methodology for the implementation of the Maintenance Plan is based on two complementary aspects: (1) Preventive conservation (material component)—undertaken directly on buildings through inspection, monitoring, maintenance and repair activities; (2) Community empowerment (intangible component)—intervening indirectly on users, through participatory and capacitation strategies, such as training on maintenance and use for local users and agents. The methodological definition envisages the formation of a multidisciplinary team [2], coordinated by an Architect and advised by a Civil Engineer, as far as the inspection, structural monitoring and risk control activities are concerned, and by a Conservation Technician with regard to the assets and applied arts (Ferreira 2015).

The Maintenance Plan has four implementation phases. A first phase (Information) consists of the gathering of information and its integrated management through the creation of a computer software and web Software/app ("MPlan") developed by the author of this paper under a specific project. This computerised software makes it possible to centralise and research information, draw up maintenance plans and make budget estimates, among other functionalities. The "MPlan" App is accessible via mobile devices (smartphones and tablets) and facilitates the filling in of forms in the course of the field work, thus optimising the implementation of the Maintenance Plan by the different operators and stakeholders involved in the process.

A second phase (Programmatic) consists in drawing up the Maintenance Plans, which are divided into five sections: Identification, Characterization, Diagnosis, Maintenance, Utilization (Figure 5). The Identification section (I) includes general information on the Building (type, chronology, description, location) and the Users (contacts, age, function, responsibility). The Characterization section (II) contains the Characterization Form (materials, techniques, durability, areas), Building's Drawings and a Chronology with building's previous interventions. The Diagnosis section (III) comprises the Anomalies Form (description, mechanism, possible causes, percentage, risk factor), the Utilization Form (including uses and related risks) and the Log of Occurrences reported by users. This section allows also for the uploading of different kind of inspection and NDT analysis, among others: thermo-hygrometric or structural monitoring, decay mapping, thermographic captions or inspection reports. The Maintenance section (IV) relates to the scheduling and description of inspection, maintenance and repairs (Maintenance Manual) including cost estimate which is a very useful tool for managers. This estimate is obtained through the relation between area (m^2) and costs of maintenance actions per m^2 supported on database provided by experienced contractor; the Intervention Form relates to specific conservation works. Finally, the Utilization section (V) provides for different kinds Manuals with recommendations for Users such as Inspection, Cleaning, Use and Current Maintenance. The Users'

Figure 5. Menu of the "MPlan" Software (Source: The author).

Manuals are documents designed for the managers and users who enjoy a continuing relationship with the building, containing information and recommendations about its correct use, preventive actions, routine maintenance or cleaning tasks, as well as small-scale maintenance and repair work. These documents are written in an accessible language and are separated into two versions: written schedules for the managers and illustrated version for the everyday users, consisting of an Instruction Manual or Good Practices Manual. The sections mentioned above compose the Maintenance Plan for each building, which can be printed either in part or in full. Many of these forms have precompiled or dropdown fields in order to facilitate their full and rapid completion, as well as to reduce errors. There is a detailed tutorial to help people to fill in the form with lists and descriptions (materials, constructive systems, damage atlas, repair actions, users' recommendations, etc). The software has reserved access through different logins (Administration, Manager, Technician, User) and provides for a system of notifications and alerts related to occurrence's management directed to specific users.

A third phase (Execution) concentrate on the execution of different maintenance works according to the Maintenance Plan's schedule (roofs, walls, pavements, doors, etc). Moreover, this phase previews two different models of training workshops: (1) Technical Workshops for local contractors and municipal maintenance teams, focusing on traditional construction skills, minor repairs, health and safety, etc. (2) Good Practices

Workshops for regular users, such as managers, priests, caretakers, cleaning and maintenance technicians, focusing on correct use and cleaning procedures, reporting occurrences, behaviour in emergency situations, etc.

Finally, since the Maintenance Plan is a dynamic instrument relating to a work in progress, a final phase (Evaluation) relates to the assessment of results and the updating of documents.

3.3 *Implementation of the maintenance plan*

The Maintenance Plan of the Monuments of the Rota do Românico is a form of exploratory applied research, currently in its implementation phase, benefiting from structural funding (FEDER) from the European Union (85%) and from the Municipal Councils (15%).

In a first stage (2014–2015), the implementation of the Maintenance Plan was tested at 12 pilot monuments (Figure 3), one of each type identified in the Vale do Sousa and Baixo Tâmega regions (monastery, church, tower, bridge, memorial and castle) with the objective of better evaluating procedures and estimating costs. In selecting the buildings, a geographical criterion was used, namely that each municipality should have only one pilot monument, in order to distribute them evenly across the territory [3].

Maintenance Plans were drawn up for the pilot monuments, through *in situ* surveys and inspections and a detailed compilation made through the use of the "MPlan" App, in accordance with the above-mentioned sections: Identification, Characterisation, Diagnosis, Maintenance and Utilisation. The Maintenance Plans drawn up for the pilot monuments supported the definition of lists of the works to be performed at each of the 12 monuments, involving a total amount of €30.000 (the works executed by an experienced contractor in 2015). Among others these included the following actions (applied case by case in accordance with each Plan): clearance of the infesting vegetation in the surrounding area, cleaning of paths providing access to the building, inspection and maintenance of the roofs (with the replacement of broken tiles and the correction of specific problems), cleaning of rainwater goods, removal of biological colonisations on the walls, replacement of broken window panes, treatment and painting of doors and windows, disinfestation of altars, maintenance of the electrical installations, correction of specific problems, etc. It is also important to notice that these maintenance works did not affect the architectural design project and aimed to re-establish the proper functioning of the buildings, while preserving their integrity and authenticity.

As far as improving the conditions for future maintenance is concerned, it was found that, in most cases, the installation of lifelines in roofs not only had a negative visual impact, but was also not economically advantageous, due to the cost of obtaining the compulsory annual safety certificates, when compared with the cost of renting mobile platforms and removable systems for the annual maintenance of roofs. Other possibilities such as rope climbing specialists (i.e. Monumentenwacht) or GutterClear scheme in England (Vandesande, 2017, 84–85) would require previous specific training and organization.

Moreover, training workshops were carried out together with the execution of maintenance works: a Technical Workshop for local contractors (Church of Telões, 17th June 2015) with "hands on" demonstrations on roofs, walls and door inspection, treatment and repair; a Good Practices Workshop for caretakers (Church of Telões, 17th September 2015) with practical simulations of ckecking, cleaning and correct use. These workshops included the presentation of illustrated Instruction Manuals for everyday users (Figure 6) and a Quiz for children, both of which highlighted the main problems of a lack of maintenance at a religious building. These are extremely important instruments, since they acknowledge that the Maintenance Plan will only be truly sustainable with the qualification and empowerment of the present and future guardians of this heritage.

In a second phase (2017–2018) the Maintenance Plan is now being extended to the 58 monuments that constitute the Rota do Românico. This stage envisages the improvement of the database and computer software ("MPlan"), providing for the inclusion of risk assessment and the inventorying of assets and plans for the evacuation of these pieces in emergency situations. Furthermore, works and training activities are being carried out at the monuments, with a total budget of €260.000. As far as resources are concerned, this amount includes human resources, such as multidiscipli-

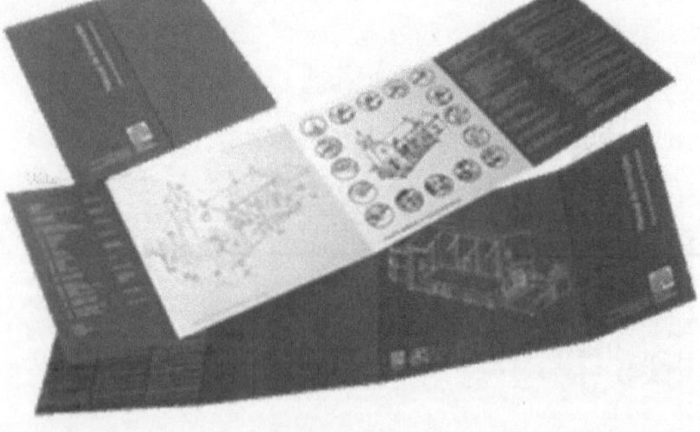

Figure 6. Illustrated Instruction Manual for a Religious Building (Source: The author).

nary teams responsible for the coordination and development of maintenance plans, inspection and training activities, as well as works being undertaken on roughly 40 monuments, performed by an experienced contractor.

3.4 *Final note and current challenges*

The Rota do Românico is a successful experience in the bridging of planned conservation and community empowerment. Hence, it demonstrates several benefits in the implementation of preventive conservation strategies, such as the reduction of costs (over the medium and long term), the preservation of heritage authenticity (increasing its cultural and tourism values, and consequently its economic value), the creation of employment (small local contractors), and, finally, the improved qualification of local agents and users for the future safeguarding and sustainability of their heritage.

The main challenges currently being faced by the project include such issues as the correct balance between maintaining an approach that is "simple and operative" (quickly and effectively applicable to a large number of buildings) and one that is simultaneously "rigorous and flexible" (adjusting to the complexity and specificity of each building). Other emerging topics include adapting the structure of the Maintenance Plans to more complex buildings, as well as improving the technology used in the survey, inspection, diagnosis and management of the buildings (drones, monitoring sensors, photogrammetry, portable laser scanners, 360° digital cameras, Heritage BIM, among others). Another issue relates to the future management model, agreed between the various agents and parties involved in the process (Rota do Românico, Municipalities, Dioceses, Regional Directorate of Culture, Tourism of Portugal, Owners, Sponsors, Users, others).

Finally, it is important to note that the project is still in its implementation phase and retains a critical stance in regard to future developments.

On the other hand, future research can possibly consider the relationship between values and threats as a tool aiming to prioritize the actions and the investments, in order to enable a long-term sustainable policy to implement planned-preventive conservation of monuments.

4 CONCLUSIONS

The Rota do Românico is a successful experience bridging planned conservation and community empowerment. Hence, it demonstrates several benefits of the implementation of preventive conservation strategies such as the reduction of costs (in medium and long-term period), the preservation of heritage authenticity (increasing the cultural and tourism values, and consequently the economic value), the creation of employment (small local contractors), and finally, the involvement and qualification of local agents and users for the future safeguarding and sustainability of their heritage.

In Portugal there are still very few instances of the implementation of preventive-planned conservation strategies. There are many possible reasons why maintenance is not carried out: firstly, because it is a cultural problem, related to the loss of maintenance habits, especially after the industrialisation of construction; and also because it involves discreet interventions with little media impact and, normally, politicians, investors and citizens prefer highly visible interventions. On the other hand, these kinds of ordinary interventions can also be less appealing for technicians because they are anonymous. Moreover, it is important to note that the economic advantages of maintenance run counter to the immediate interests of the professionals involved (contractors, architects and engineers earn a percentage of the total cost). Some other causes related to the failure to implement maintenance practices are the lack of skilled labour (in many cases aging and in risk of disappearing), as well as the difficulty in applying or gaining access to funding for these kinds of strategies and, finally, the absence of a suitable legal and institutional framework regarding preventive-planned conservation and maintenance practices.

The main contribution of this paper is to put into perspective the sustained process of applying the concepts of preventive-planned conservation linked to community participation, in diverse experiments that have common conceptual and methodological frameworks, but different specificities and conditions of application. Furthermore, the paper emphasises the role of society in the execution of preventive conservation measures and the consequent responsibility for these, which can have an important impact, not only from the economic point of view, but also from the perspective of social empowerment and the formulation of a constant relationship of greater benefits for both society and the heritage.

The Rota do Românico project provides a clear and operative model for adopting a methodological approach (1) Planned conservation (performed directly on buildings) and (2) Community empowerment (undertaken indirectly on users), a phasing of its implementation (Information, Programming, Execution, Evaluation) and a structure for Maintenance Plans (Identification, Characterisation, Diagnosis, Maintenance, Utilisation). All of these are complemented with illustrated training manuals, while also being supported by computer software/App and including assessment on economic data. Hence, this model, bridging preventive-planned conservation with community empowerment, can be possibly applied and adapted to other types of case studies.

In this way, the cases presented in this paper are a valuable resource both for research and for the professional field, and the proposed model may therefore help in identifying possible paths for preventive-planned conservation and community empowerment for the built heritage in other parts of the world.

NOTES

1. With a budget of €395,062.95 the work took place between June 2016 and April 2017. Coordination of Eliseu Gonçalves with support of José Luís Gomes, under the guidance of Álvaro Siza.
2. Management Plan for the Conservation and Valorisation of the Monuments of the Romanesque Route—Concept and Coordination: Prof. Arch. Teresa Cunha Ferreira; Collaboration: Arch. Margarida Ramos, Joana Gonçalves, Paulo Mendes (2014/15); Marta Ferreira (2016/2017); Advisory: Ing. Esmeralda Paupério, Dr. Jaime Duarte, Ing. Filipe Ferreira (AOF); Management: Dr. Rosário Machado; Ing. Ricardo Magalhães, Nelson Antunes, Vítor Marinho.
3. Pilot monuments: Monastery of São Pedro Ferreira; Church of São Cristovão Nogueira; Bridge of Arco; Church of São Mamede Vila Verde; Bridge of Espindo (Lousada); Castle of Arnoia; Church of São Gens de Boelhe; Monastery of Cárquere (Resende); Church of Telões (Amarante); Tower of Aguiar de Sousa (Paredes); Monastery of Santo André de Ancede; Memorial of Sobrado.

AKNOWLEDGEMENTS

This research is co-financed by the European Regional Development Fund (ERDF) through the COMPETE 2020 – Operational Programme Competitiveness and Internationalization (POCI) and national funds by the FCT under the POCI-01-0145-FEDER-007744 and FCT IF/01501/2014. The author kindly aknowledges the Rota do Românico, Faculty of Architecture of the University of Porto and ICOMOS Portugal.

REFERENCES

Calejo, R. (2009). *Manutenção de Edifícios*. Porto: FEUP.

Cóias, V. (2004). Guia prático para a conservação de imóveis: manual para a utilização durável e económica da habitação, através de uma adequada manutenção. Lisboa: Dom Quixote.

Costa, A.A. (2003). Faculdade de Arquitectura da Universidade do Porto. In FAUP Publicações (ed.) *Edifício da Faculdade de Arquitectura da Universidade do Porto Percursos do Projecto*, Porto: FAUP Publicações.

Crofts, S. 2008. The Good Maintenance Guide- A Practical Handbook To Help Volunteers To Care & Preserve Our Historic Places Of Worship, London: S.P.A.B.

Della Torre, S. et al. (2003). La conservazione programmata del patrimonio storico architettonico—Linee guida per il piano di manutenzione e il consuntivo scientifico. Milano: Guerini e Associati.

Ferreira, T.C. (2011ª). Manutenção e Património corrente. Conceitos e perspectivas. *Actas do Seminário Cuidar das casas: A Manutenção do Património Corrente*, ICOMOS-Portugal/FEUP, Porto, pp. 1–8.

Ferreira, T.C. (2011b). 'Rota do Romanico': un processo innovativo nel contesto portoghese, Atti del XXVI Convegno Scienza e Beni Culturali (Bressanone) - Governare l'innovazione: processi, strutture, materiali e tecnologie tra passato e futuro, Venezia: Arcadia Ricerche.

Ferreira, T.C. (2012). Prevenção e Manutenção. Conceitos e aplicações. Estudo de caso: a Rota do Românico no Norte de Portugal, *4º Congresso de Patología y Rehabilitación de Edificios– Patorreb* 2–14 de Abril, FEUP, Porto, 2012, p. 6.

Ferreira, T.C. (2015). Manutenção de edifícios com valor patrimonial. O caso do Plano de Manutenção dos Monumentos da Rota do Românico, *Simpósio Manutenção De Edifícios: Perspetiva Multidisciplinar (GEQUALTEC)*, FEUP, Porto, pp. 23–30.

Ferreira, T.C. (2016). Faculdade de Arquitectura da Universidade do Porto. Perspectivas para a manutenção, in Ferreira, T.C. e Rocha, P.F., Saber manter os edifícios: pensar, desenhar, construir, CEAU-FAUP/CEES-FEUP, Porto.

Ferreira, T.C. (2017). Contributions for the implementation of preventive conservation and maintenance strategies in the Faculty of Architecture of the University of Porto, *REHAB 2017–3rd International Conference on Preservation, Maintenance and Rehabilitation of Historical Buildings and Structures*, 14–16 July 2017. Braga.

Flores Collen. I. & Brito, J. (2003). Planos de Manutenção Pro-activa em Edifícios recentes, *3º Encore - Encontro sobre Conservação e Reabilitação de Edifícios*. Lisboa, LNEC, pp. 1–10.

Forster, M.A. & Kayan, B. (2009). Maintenance for historic buildings: a current perspective. Structural Survey. Vol. 27. Issue number 3: pp. 210–229.

Henriques, F. (1991). A conservação do património histórico edificado. Lisboa: LNEC.

Heras, V., Wijjfels, A., Cardoso, F., Vandesande, A., Santana, M., Orshoven, J.V., Steenberghen, T. and Koenraad, V.B. (2013). A value-based monitoring system to support heritage conservation planning", Journal of Cultural Heritage Management and Sustainable Development.Vol.3. Issue Number 2: pp. 130–147.

ICOMOS. (2011). Paris Declaration on heritage as a driver of development, Paris: UNESCO.

ICOMOS. (2014). The Florence Declaration on Heritage and Landscape as Human Values. Florence: UNESCO.

Lipovec, N. & Van Ballen, K. (2010). Tra prevenzione e manutenzione: i "Monumentenwatchen". In Biscontin, G. & Driussi, G. (eds), Atii del XXVI convegno internazionale Scienza e Beni Culturali: 193–202. Venice: Arcadie richerche.

Paiva, J., Aguiar, J. and Pinho, A. (2006). *Guia Técnico de Reabilitação Habitacional*. 2nd vol. Lisboa: INH/LNEC.

Ramos, L.F., Morais, M, Azenha, M., Masciotta, G., Pereira, E., Ferreira, T.C. & Lourenço, P.B. 2017. Monitoring and Preventive Conservation of the Historical and Cultural Heritage: the HeritageCare Project, *Congresso da Reabilitação do Património (CREPAT)* 29–30 June 2017, Universidade de Aveiro.

Roders, A. (2007). *Re-architecture*, Phd thesis, Eindhoven University of Technology.

Romão, X., Paupério, E., Ferreira, T., Vasconcelos, D. & Rio, F. (2013). Janelas com tinta têm muita + pinta. [online Video]. 7 Fevereiro 2013. Available from: https://www.youtube.com/watch?v = OO3nBLXlbnU [Accessed on: 29 de Maio 2017].

Sacco, P.L. (2009). *Distretti culturali come shock cultural*, in Putignano, F. (coord.) *Learning Districts*, Maggioli, Milano, pp. 47–60.

Silva, D. (2016). *Perspectivas para a manutenção e utilização do Pavilhão Carlos Ramos na FAUP*, Master's Dissertation, Faculdade de Arquitectura da Universidade do Porto.

Siza, A. (2016). Interview by T.C. Ferreira. In: Ferreira, T.C. and Rocha, P.F. (2016), *Saber manter os edificios: pensar, desenhar, construir*, CEAU-FAUP/CEES-FEUP, Porto.

Van Balen, K. & Stulens. A. (2010). Current Preventive Conservation Approach: the Precomos Unesco Chair's perspective, paper presented at UNESCO Chair on Preventive Conservation, Maintenance and Monitoring of Monuments and Sites (PRECOM²OS), 20 September 2010. Belgium: Leuven.

Van Balen, K. & Vandesande, A. (Eds.) (2013). Reflections on Preventive Conservation, Maintenace and Monitoring of Monuments and Sites. Leuven: Acco.

Vandesande, A. & Van Balen, K. (2016). An operational preventive conservation system based on the Monumentenwacht mode. In: Van Balen, K. & Verstrynge, E. (eds.) SAHC 2016 Structural Analysis of Historic Constructions. Anamnesis, diagnosis, therapy, controls. Leiden: (CRC Balkema / Taylor & Francis Group.

Vandesande, A. (2017). Preventive conservation strategy for built heritage aimed at sustainable management and local development, PhD Dissertation, KU Leuven University: Faculty of Engineering Science.

Innovative Built Heritage Models – Van Balen & Vandesande (Eds)
© 2018 Taylor & Francis Group, London, ISBN 978-1-138-49861-7

Author index